Springer Series in
CLUSTER PHYSICS

Springer
*Berlin
Heidelberg
New York
Barcelona
Hong Kong
London
Milan
Paris
Singapore
Tokyo*

Springer Series in
CLUSTER PHYSICS

Series Editors:
A. W. Castleman, Jr.   R. S. Berry   H. Haberland   J. Jortner   T. Kondow

The intent of the Springer Series in Cluster Physics is to provide systematic information on developments in this rapidly expanding field of physics. In comprehensive books prepared by leading scholars, the current state-of-the-art in theory and experiment in cluster physics is presented.

Mesoscopic Materials and Clusters
Their Physical and Chemical Properties
Editors: T. Arai, K. Mihama, K. Yamamoto and S. Sugano

Cluster Beam Synthesis of Nanostructured Materials
By P. Milani and S. Iannotta

Theory of Atomic and Molecular Clusters
With a Glimpse at Experiments
Editor: J. Jellinek

Dynamics of Clusters and Thin Films on Crystal Surfaces
By G. Rosenfeld

P. Milani   S. Iannotta

Cluster Beam Synthesis of Nanostructured Materials

With 118 Figures and 9 Tables

Springer

Dr. Paolo Milani
INFM-Dipartimento di Fisica
Università di Milano
Via Celoria 16
20133 Milano
Italy

Dr. Salvatore Iannotta
Centro di Fisica
degli Stati Aggregati-CNR/ITC
Via Sommarive 18
38050 Povo di Trento
Italy

Series Editors:

Professor A. W. Castleman, Jr.
(*Editor-in-Chief*)
Department of Chemistry
The Pennsylvania State University
152 Davey Laboratory
University Park, PA 16802, USA

Professor R. Stephen Berry
Department of Chemistry
The University of Chicago
5735 South Ellis Avenue
Chicago, IL 60637, USA

Professor Dr. Hellmut Haberland
Albert-Ludwigs-Universität Freiburg
Fakultät für Physik
Hermann-Herder-Strasse 3
D-79104 Freiburg, Germany

Professor Dr. Joshua Jortner
School of Chemistry, Tel Aviv University
Raymond and Beverly Sackler
Faculty of Sciences
Ramat Aviv, Tel Aviv 69978, Israel

Dr. Tamotsu Kondow
Toyota Technological Institute
Cluster Research Laboratory
East Tokyo Laboratory
Genesis Research Institute Inc.
Futamata 717-86
Ichikawa, Chiba 272-0001, Japan

ISSN 1437-0395
ISBN 3-540-64370-2 Springer-Verlag Berlin Heidelberg New York

Library of Congress Cataloging-in-Publication Data

Milani, Paolo, 1960– Cluster beam synthesis of nanostructured materials / Paolo Milani, Salvatore Iannotta. p. cm. – (Springer series in cluster physics; 1) Includes bibliographical references and index. ISBN 3-540-64370-2 (alk. paper) 1. Nanostructure materials. 2. Ion bombardment. 3. Nanotechnology. I. Iannotta, Salvatore, 1953– . II. Title. III. Series. TA418.9.N35M55 1999 620.1'1–dc21 98-52462

This work is subject to copyright. All rights are reserved, whether the whole or part of the material is concerned, specifically the rights of translation, reprinting, reuse of illustrations, recitation, broadcasting, reproduction on microfilm or in any other way, and storage in data banks. Duplication of this publication or parts thereof is permitted only under the provisions of the German Copyright Law of September 9, 1965, in its current version, and permission for use must always be obtained from Springer-Verlag. Violations are liable for prosecution under the German Copyright Law.

© Springer-Verlag Berlin Heidelberg 1999
Printed in Germany

The use of general descriptive names, registered names, trademarks, etc. in this publication does not imply, even in the absence of a specific statement, that such names are exempt from the relevant protective laws and regulations and therefore free for general use.

Typesetting by the author using a Springer TEX macro package
Cover concept: eStudio Calamar Steinen
Cover production: *design & production* GmbH, Heidelberg
SPIN: 10647008   57/3144/mf - 5 4 3 2 1 0 – Printed on acid-free paper

Preface

On appelle Science l'ensemble des recettes qui réussissent toujours,
et tout le reste est littérature. *Paul Valéry*

One of the most used and abused combining forms in the past few years in solid-state physics and materials science is "nano". The "nanoworld" as one can read also in newspapers and magazines, is populated by nanoparticles, nanocrystals, nanotechnologies, nanomachines, nanocomputers, and so on.

Among many enthusiastic prophecies on the "nanofuture" we feel that a few words of caution are in order. The well established semiconductor-based technology is rapidly reaching its ultimate limits in terms of miniaturization down to dimensions of a few tens of nanometers. In this sense, nanoelectronics is already a reality and many efforts are addressing the goal of breaking the dimension limit imposed by the present technology. From a more general point of view the applications of nanomaterials, i.e. materials composed of elementary building blocks of nanometer size, are still a matter of speculation and it is presently difficult to extrapolate the future industrial applications.

Production and manipulation of clusters in many cases is still an art, not yet a science, and the exciting results obtained in many laboratories are waiting for new technological solutions to be reproduced on a large, economically convenient, scale. The efforts towards the development of efficient production, manipulation and characterization methods in view of possible applications should not lead one to overlook the fact that clusters and cluster-assembled materials are very interesting from a fundamental point of view and that a basic comprehension of these systems is a necessary background for the development of any cluster-based technology.

This book is devoted to cluster production and manipulation methods based on molecular beams. This is a technique rich in opportunities for both pure and applied science. Recently, cluster beams have begun to be considered not only as an academic subject, but also as interesting candidates for nanostructured material synthesis and processing, in conjunction with or as an alternative to more traditional ion beams. This change of perspective has been stimulated by the development of new cluster beam technologies.

This book has the aim of presenting in a coherent and synthetic way the basic principles of cluster beams and of related techniques. This should provide a solid basis for a critical evaluation of advantages, drawbacks and perspectives of cluster beams for assembling nanostructured materials. We also emphasize those interdisciplinary aspects relevant for coupling cluster beams with other growth methods and surface science methodologies. In this spirit, this book is aimed at scientists working with clusters, nanostructured materials and thin films synthesis, as well as students entering the field of clusters. We also hope that this book will represent a starting point for further advances and systematic achievements in the field of cluster-assembled materials.

The authors wish to thank the many people who contributed to the realization of this work. First of all we thank our families for their invaluable support. We gratefully acknowledge for their help, advice and encouragement G. Benedek, P. Piseri, E. Barborini, C.E. Bottani, A. Li Bassi, A. Ferrari, F. Biasioli, G. Ciullo, M. van Opbergen, C. Corradi, M. Mazzola, A. Boschetti, M. Sancrotti, L. Colombo, C.E. Ascheron, R.A. Broglia. We are indebted to I. Yamada, G. Takaoka, T. Takagi, K. Sumiyama, H. Haberland, K.-H. Meiwes-Broer, A. Perez, for providing results prior to publication and for many helpful discussions. Financial support from INFM (Advanced Research Project CLASS), CNR (Projects FILINCLUBE, Industrial Applications of Plasma, MADESS II and PFMAT II) and INFN (Project Cluster) is acknowledged.

Milan, Trento,
January 1999 *Paolo Milani, Salvatore Iannotta*

Contents

1. **Introduction** 1
2. **Molecular Beams and Cluster Nucleation** 5
   2.1 Molecular Beams 5
      2.1.1 Continuous Effusive Beams 6
      2.1.2 Continuous Supersonic Beams 9
      2.1.3 Pulsed Beams 18
   2.2 Nucleation and Aggregation Processes 20
      2.2.1 Classical Theory 20
      2.2.2 Homogeneous Nucleation by Monomer Addition 21
      2.2.3 Homogeneous Nucleation by Aggregation 26
      2.2.4 Nucleation of Clusters in Beams 27
      2.2.5 Semi-empirical Approach to Clustering in Free Jets ... 30
3. **Cluster Sources** 35
   3.1 Vaporization Methods 35
      3.1.1 Joule Heating 36
      3.1.2 Plasma Generation for Cluster Production 42
      3.1.3 Laser Vaporization 42
      3.1.4 Glow and Arc Discharges 47
   3.2 Continuous Sources 52
      3.2.1 Effusive Joule-Heated Gas Aggregation Sources 52
      3.2.2 Magnetron Plasma Sources 61
      3.2.3 Supersonic Sources 62
   3.3 Pulsed Sources 70
      3.3.1 Pulsed Valves 70
      3.3.2 Laser Vaporization Sources 76
      3.3.3 Arc Pulsed Sources 83
4. **Characterization and Manipulation of Cluster Beams** 91
   4.1 Mass Spectrometry 91
      4.1.1 Quadrupole Mass Spectrometry 91
      4.1.2 Time-of-Flight Mass Spectrometry 96
      4.1.3 Retarding Potential Mass Spectrometry 107

    4.2 Detection Methods 109
        4.2.1 Ionization of Clusters 109
        4.2.2 Charged Cluster Detection 110
        4.2.3 Cluster Beam Characterization 117
    4.3 Cluster Selection and Manipulation 120
        4.3.1 Size and Energy Selection 120
        4.3.2 Quadrupole Filter 122
        4.3.3 Separation of Gas Mixtures in Supersonic Beams 123

5. **Thin Film Deposition and Surface Modification by Cluster Beams** 125
    5.1 Kinetic Energy Regimes 128
    5.2 Diffusion and Coalescence of Clusters on Surfaces 133
    5.3 Low-Energy Deposition 142
        5.3.1 Cluster Networks and Porous Films 143
        5.3.2 Composite Nanocrystalline Materials 153
    5.4 High-Energy Deposition 155
        5.4.1 Implantation, Sputtering, Etching 155
        5.4.2 Thin Film Formation 161

6. **Outlook and Perspectives** 167
    6.1 Cluster Beam Processing of Surfaces 168
    6.2 Nanostructured Materials Synthesis 171
    6.3 Perspectives 172

Appendix 175

References 179

1. Introduction

The properties of clusters depend on the number of their constituents: by controlling the size of a cluster one can change the electronic, optical, structural and chemical characteristics of an aggregate. Moreover, by assembling clusters, one can produce a material with novel functional and structural properties. This is easy to say but difficult to do: among different aspects that characterize clusters, high reactivity can be considered universal. The interaction and coalescence of aggregates deposited on a substrate are largely unknown so that one can wonder to what extent clusters retain their individuality in cluster-assembled materials. Another major problem is the production of large quantities of particles with well defined properties (number of components, structure, etc.).

The understanding of the fundamental properties of clusters and the possibility of using this class of objects for technology is intimately connected to the availability of production, manipulation and characterization methods capable to preserve the individuality of the clusters while controlling their dimensions and structures. If devising synthetic routes is a formidable challenge towards the applications of clusters, it should not be undervaluated the importance of finding diagnostic techniques capable to characterize clusters and cluster-assembled materials over the whole range of length scales in which the precursor aggregates organize themselves. Different excitations (plasmons, optical and acoustic phonons) and properties are sensitive to different scale and hierarchies of organization.

Many physical and chemical routes are currently being used for the synthesis of clusters and for assembling nanostructured materials, and it is universally accepted that a unique approach for producing clusters with well controlled properties does not exist. A particular synthetic method is effective depending on the physical-chemical properties of each element or material which, in turn, are very often different from the bulk and mostly unknown.

The general requisites that one should look for in assembling clusters are: the control on mass distribution, structure and chemical reactivity. Moreover one should be able to control the degree of coalescence of the clusters during the formation of the nanostructured material. It is then necessary to adopt diagnostic techniques that adequately monitor every step of both the cluster formation and the cluster assembling processes.

1. Introduction

This book is devoted to the production of clusters in molecular beams and to their use for the synthesis and processing of nanostructured materials. The field of molecular beams has been particularly fruitful since the first applications dating back to the twenties, for both pure an applied science. If the impact of molecular beam epitaxy (MBE) on science and technology does not need even to be reminded, the contribution of supersonic molecular beams to physics and chemistry is also conspicuous. Nevertheless, with the remarkable exception of MBE, molecular beam techniques seems to belong to laboratories where fundamental research is conducted, without definite perspectives for technological applications. Cluster beams share the same destiny: supersonic cluster beams provide the opportunity of investigating the fundamental properties of aggregates free of any contamination and substrate interaction, with a very high degree of control on cluster size. However, the limited quantity of clusters that can be produced and the complexity of the apparatus required for cluster beams, made this technique a weak candidate for applications.

The properties of free clusters in molecular beams have been studied with many different techniques, giving rise to a wide range of results described and discussed in several review papers, monographs and conference proceedings. For a wide-ranging overview, the reader is referred to the following sources [1.1–1.3].

Only in the last few years, we have assisted to a growing interest for applications of cluster beams. It is motivated by the recent development and improvement of several production and manipulation techniques capable of fulfilling the requirements for surface processing and material synthesis where ion beams have almost reached their intrinsic ultimate limits.

We conceived this book with the aim of providing a synthesis of the various molecular beam methods suitable for producing clusters. Moreover we aimed at discussing systematically and coherently the wealth of experimental results that have been accumulated on the production and application of clusters beams for the synthesis of nanostructured materials. This monograph is intended as a working instruments for physicists, chemists and material scientists interested in molecular beam techniques and their application to clusters. We discuss the present state of the art and describe critically those experimental aspects that could realistically produce an impact in the use of cluster beams as a reasonable and useful tool in different fields of materials science.

We will show that to this end, the interdisciplinary character, which is inherent to this field, must even develop further with the introduction of techniques and procedures typical of surface science and solid state physics.

The monograph is organized in the following way: after this introduction, Chap. 2 deals with the basic principles of effusive and supersonic molecular beams with particular attention to cluster formation. Different beam regimes and nucleation processes are described in detail. Chapter 3 illustrates and discusses the processes underlying the vaporization methods, first step for the condensation of clusters. A description of the principal cluster sources and of their operation principles follows. In Chap. 4 the basics of characterization and manipulation of cluster beams are described with particular attention on mass spectrometry methods. Chapter 5 describes and critically discusses applications of cluster beams in term of surface processing and nanostructured material assembly. Finally, Chap. 6 gives an outlook and discusses future prospects.

2. Molecular Beams and Cluster Nucleation

2.1 Molecular Beams

For people working with molecular beams, the appearance of clusters due to condensation effects is detrimental since it limits the low temperature and high intensity regimes attainable with supersonic nozzle beams. Many studies have been devoted to the characterization of condensation processes in order to avoid cluster formation. On the other hand, if one is interested in the clusters themselves, molecular beams are a very rich area for the characterization of their fundamental properties [2.1, 2.2]. There are several basic reasons to use molecular beams for the production and investigation of clusters: primarily the control on temperature and pressure of the gas forming the beam can produce the conditions of an efficient condensation. Moreover, clusters in molecular beams can be studied without the interference of matrices and/or substrates; their kinetic energy, in the eV range, is very interesting for a wealth of physico-chemical phenomena; their mass distribution can be controlled and they can be mass selected.

The typical apparatus to produce and to study cluster and molecular beams is basically the same. Source geometry and operating conditions determine the degree of condensation and hence the presence of aggregates in the beam. In general cluster beams will require a more complex apparatus with large vacuum pumps and parts maintained at very high temperatures. As shown in Fig. 2.1 a cluster beam apparatus consists of a cluster source maintained at a stable pressure P_0 and temperature T_0 that can be varied, the beam extraction system made of several skimmers and/or collimators, and vacuum pumps to maintain low pressures in the various chambers. As we will see, the type of sources and the operating regimes can be very different, however it is possible to identify some basic characteristics common to both cluster as well as molecular beam apparatus.

Molecular beams can be considered as a powerful tool for the preparation of clusters and for their potential technological use. Due to the close connection between beams of clusters and molecular beams, we will present and discuss in this chapter some basic concepts related to molecular beams with the aim of giving the reader a general framework and a useful background. The different topics will be only summarized rather than give an exhaustive

6   2. Molecular Beams and Cluster Nucleation

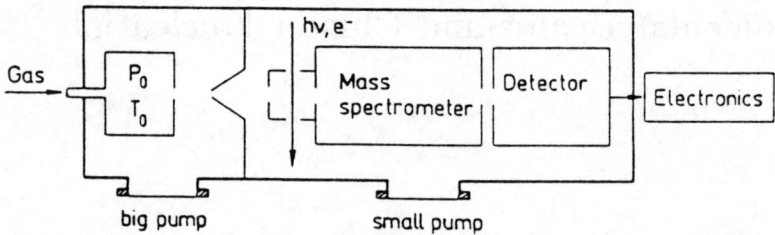

Fig. 2.1. Scheme of a typical molecular beam apparatus. The two major constituent parts are the source chamber and the characterization chamber, separated by a collimator (skimmer). The beam in the second chamber is ionized and then analyzed by a mass spectrometer. (From [2.2])

treatment that is beyond the scope of the present book. A thorough and comprehensive treatment of molecular beam methods is already available [2.3].

2.1.1 Continuous Effusive Beams

The basic concept of an effusive beam source is very simple: it is an orifice in a very thin wall of a reservoir where the gas or vapor is in thermal equilibrium. The opening is small enough so that the outgoing flow of atoms or molecules will not affect the equilibrium into the reservoir. If the pressure in the reservoir is low enough, the outgoing flow will be molecular so that the effusion rate and both the angular and velocity distributions of the formed beam can be calculated on the basis of gas kinetic theory without any assumption. Thermal equilibrium and molecular flow give rise to a beam of atoms and/or molecules that has well defined equilibrium distributions of internal states which can be controlled by changing the temperature of the reservoir.

A purely effusive source is typically based on a suitable combination of a furnace and an opening of diameter d. It operates in a regime where the Knudsen number $K_n = \lambda/d > 1$ so that the atoms effuse out of the source. The parameter that controls the source regime is the product $P_0 d$: a vapor pressure lower than 2×10^{-1} mbar and oven openings up to about 2 mm in diameter give rise to a molecular flow. The mean free path of the atoms of a vapor at the pressure P_0 (in Torr) in a furnace at a temperature T_0 (in Kelvin) is given by [2.4]

$$\lambda = \frac{7.321 \times 10^{-20} T_0}{P_0 \sigma} \text{cm} , \qquad (2.1)$$

where σ is the collision cross-section in cm^2. The atoms or molecules effuse out of the source with a cosine angular distribution for apertures that have walls thin compared to the diameter. The typical intensity into a solid angle $d\omega = \sin\theta \, d\theta \, d\phi$, at an angle θ with respect to the normal to the orifice, is then given by:

$$dN = \frac{d\omega}{4\pi} n_0 \bar{v} A \cos\theta = \chi A \cos\theta \, \frac{d\omega}{\pi} \,, \tag{2.2}$$

where n_0 is the number density in the source in atoms cm^{-3}, $\bar{v}$ is the average thermal molecular velocity in cm s^{-1} of the particles in the oven (Maxwell–Boltzmann distribution)

$$\bar{v} = 1.4551 \times 10^4 \sqrt{\frac{T_0}{M}} \,, \tag{2.3}$$

where A is the area of the orifice and M is the mass in atomic units. The parameter

$$\chi = n_0 \bar{v}/4 \tag{2.4}$$

gives the rate at which the atoms or molecules in the reservoir cross a unit area of the orifice so that the total flux rate of the source is χA. The forward intensity ($\theta = 0$) expressed in particles sr^{-1} s^{-1} is then given by (2.2):

$$I(0) = \frac{\chi A}{\pi} = 1.125 \times 10^{22} A \frac{P_0}{\sqrt{MT_0}} \,. \tag{2.5}$$

The normalized velocity distribution of an effusive beam is given by the Maxwellian distribution

$$f(v)dv = \frac{4}{\sqrt{\pi}} v^2 e^{-v^2} dv \,, \tag{2.6}$$

where $v = v/v_{\text{mp}}$ is the reduced velocity and $v_{\text{mp}} = \sqrt{2kT_0/m}$ is the most probable velocity.

Effusive sources can have circular or slit orifices that allow specific designs for deposition or spectroscopy measurements. In the case of a circular hole, one should keep in mind that the maximum intensity achievable, while maintaining a molecular flow regime, is given by the following approximated formula for the orifice area (Knudsen condition)

$$A \approx \pi \lambda \,. \tag{2.7}$$

For a given pumping system, since λ is inversely proportional to the pressure and the source diameter has to be reduced quadratically with the pressure, the maximum forward intensity decreases linearly with pressure.

Very often, instead of a thin aperture, a short tube of length L is used in order to increase the directionality of the beam. If $L \gg d$, but it is still smaller then the mean free path λ, the tube is "transparent" and the flow of the particles, although still effusive, is characterized by a much higher directionality. The angular spread is then essentially given by the diameter of the tube while both the intensity (atoms s^{-1} sr^{-1}) and the flow rates are proportional to the vapor pressure in the source.

The transmissivity of the channel W is defined by

$$I = W \chi A \,, \tag{2.8}$$

where I is the total flow rate through the channel, $A = \pi r^2$ is the area of the cross-section of the tube. W can be calculated directly [2.5] and has the following form

$$W = 1 + \frac{2}{3}(1 - 2\eta)(\beta - \sqrt{1+\beta^2}) + \frac{2}{3}(1+\eta)\beta^{-2}(1 - \sqrt{1+\beta^2}), \quad (2.9)$$

where

$$\eta = \frac{1}{2} - \frac{1}{3\beta^2}\left(\frac{1 - 2\beta^2 + (2\beta^2 - 1)(1+\beta^2)^{1/2}}{(1+\beta^2)^{1/2} - \beta^2 \sinh^{-1}(\frac{1}{\beta})}\right) \quad (2.10)$$

and

$$\beta = \frac{2r}{L}. \quad (2.11)$$

For very long channels ($\beta \to 0$), the limiting value of the transmissivity becomes $W = \frac{4}{3}\beta$, while for a very short tube ($\beta \to \infty$) the expected value $W = 1$, corresponding to an orifice in a thin wall, is found.

A very interesting feature of this kind of channeled effusive source comes from the ratio between the forward intensity and the total flow rate, this can be written, as a function of W, in the form [2.3]

$$\frac{I(0)}{I} = \frac{1}{\pi W} \quad (2.12)$$

so that one can define a parameter $k_w = \pi I(0)/I = 1/W$ called peaking factor. Since the relative forward flow rate increases linearly with $L/2r$, a channeled effusive source will give the advantage that, for a given pumping system, one would obtain a much higher beam intensity.

The regime where the length of the tube is larger than the mean free path ($L > \lambda > d$) still gives rise to an effusive flow but producing larger intensities, since it allows higher source pressures. Unfortunately, in this regime, an increase of pressure produces also an increase of the angular spread of the beam, reducing its quality. A solution to this problem are the so-called multi-channel arrays, constructed by coupling a large number of tubes usually by microfabrication techniques. The typical dimensions of the tubes are a few millimeters length and $10\,\mu$m diameter. Figure 2.2 shows a multi-channel array and a skimmer assembly.

The advantages of this solution are: high uniformity over a larger area, improved directionality and larger maximum intensities compared to a source with a single aperture at the same total gas flow. Such multichannel arrays are commercially available, made of stainless steel or silicon, and have been demonstrated to give up to a factor 20 larger maximum intensities [2.6, 2.7]. An interesting effect has been demonstrated by Lucas [2.8] and Ross [2.9] by constructing focused multichannel arrays. The idea is to assemble the tubes forming the array in such a way that they all point to the same position at the required distance where the beam is actually used. In principle, the intensity increases by a factor equal to the number of tubes forming the array. A factor

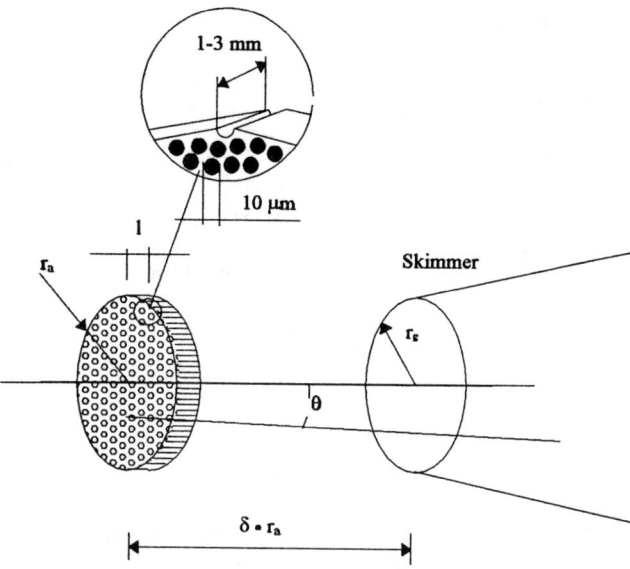

Fig. 2.2. Schematic view of a multi-channel array-skimmer assembly. In the inset the typical dimensions of one of the channels are shown

of 70 has been reported by comparing a focussed to an unfocused array for He [2.8] and a factor of 30 for Na atoms [2.9].

2.1.2 Continuous Supersonic Beams

A continuous free-jet expansion is characterized by features schematically shown in Fig. 2.3 for a short converging nozzle. In this case, the flow can be approximated by an isentropic expansion with negligible heat conduction and viscous effects.

The gas flows from the inside of the source (stagnation state), characterized by thermalized conditions (P_0, T_0), into the outside chamber at the lower pressure P_v driven by the pressure gradient $(P_0 - P_v)$. If P_v is kept low enough, the continuum flow does not prevail so that both jet boundaries and shock structures do not form. In the converging region of the nozzle, because of the reduction of the cross-section, the flow accelerates and will reach the sonic speed (Mach number $M = 1$) if

$$\frac{P_0}{P_v} > \left(\frac{\gamma+1}{2}\right)^{\gamma/(\gamma-1)} , \qquad (2.13)$$

where $\gamma = C_p/C_v$ is the ratio between the specific heats at constant pressure and volume. As P_0/P_v increases over the critical value given by (2.13), the regime changes from effusive (subsonic) to supersonic with the local pressure after the nozzle becoming $\approx P_0/2$ and therefore larger and independent of

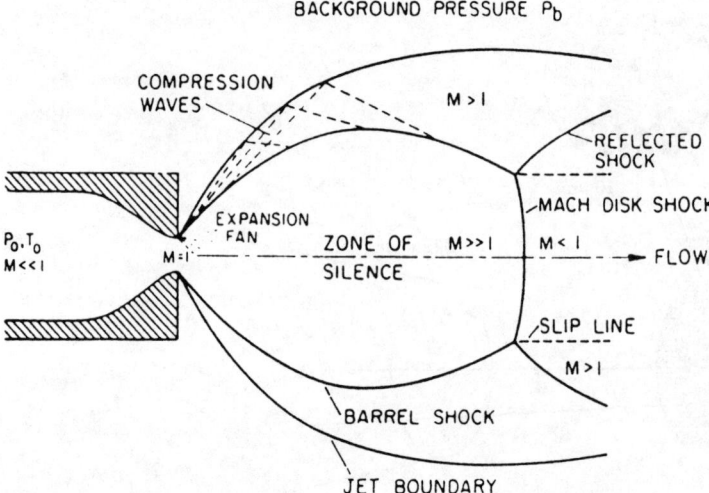

Fig. 2.3. Schematics of the features present in a continuum free-jet expansion due to the interaction between the expanding and the background gases. M indicates the Mach number. P_0, T_0 are the pressure and temperature inside the stagnation chamber while P_b is the background pressure (from [2.10])

P_v: this is why the flow is called underexpanded. At this point the gas will keep expanding to cope with the boundary vacuum conditions given by the pressure P_v. In this process, the gas keeps increasing its velocity ($M > 1$) moving faster than the speed of sound so that, to adjust to the local boundary conditions, a shock wave is formed. This consists of a narrow (of the order of the local λ), non-isentropic region characterized by high pressure, temperature and velocity gradients. The location of this region, called the Mach disk, depends only on P_0 and P_v and is placed at an experimentally determined distance

$$\delta_M = \frac{x_M}{d} = 0.67 \left(\sqrt{\frac{P_0}{P_v}} \right) . \tag{2.14}$$

The region delimited by the shock barrel and the Mach disk is the core of the expansion also called the zone of silence since the supersonic flow here does not "feel" the external conditions. Here, the molecular beam should be extracted with a least interfering conical collimator (skimmer). The interference of the skimmer, and of the wall that separates the source vacuum chamber from the rest of the apparatus (where the beam is actually used), depends strongly on the geometrical shapes and distances involved and in particular on the pressure P_v. The effect can be minimized by keeping P_v low and by properly positioning a skimmer with a conical profile and very sharp edges in the zone of silence. Electroformed skimmers are commercially available and guarantee a very good performance [2.11].

From the point of view of the beam properties, the expansion can be assumed to be isentropic and viscous and heat conduction effects can be disregarded. In this framework, the relevant quantity is enthalpy and not the internal energy since the flow work is carried out by the pressure gradient. The total enthalpy H_0 should be constant along each streamline originating from the source, so that the enthalpy per unit mass h_0 is

$$h_0 = h + \frac{v^2}{2}. \tag{2.15}$$

During the expansion, the gas cools and therefore its enthalpy h decrease, giving rise to a velocity increases. For an ideal gas, since $dh = C_p dT$,

$$v^2 = 2(h_0 - h) = 2\int_T^{T_0} C_p dT. \tag{2.16}$$

Under the assumption that C_p is constant over the range $T_0 \to T$, then $v = \sqrt{2C_p(T_0 - T)}$. If the expansion is efficient and the cooling large ($T \ll T_0$), one can determine the terminal (or maximum) velocity of the beam

$$v_\infty = \sqrt{\frac{2k}{m}\left(\frac{\gamma}{\gamma - 1}\right)T_0}, \tag{2.17}$$

where the relationship $C_p = k/m\,(\gamma/(\gamma - 1))$ for an ideal gas has been used.

It is interesting to evaluate the behavior of a mixture of gases that is often the case for cluster beams where species with different masses are present (the carrier gas and the different cluster sizes). In this case one should substitute the heat capacity and the mass with weighted averages: $\overline{C_p} = \sum_i X_i C_{p_i} = \sum_i X_i k\,(\gamma/(\gamma-1))$ where X_i is the mass fraction of the species i of mass m_i in the gas mixture. $\overline{m} = \sum_i X_i m_i$ is the average mass of the beam. $C_p = \overline{C_p}/\overline{m}$ and T_0 define the average terminal velocity of the beam and each individual species is characterized by a different energy:

$$E_i \approx \left(\frac{m_i}{\overline{m}}\right) T_0. \tag{2.18}$$

As a consequence, in the expansion, a heavy species is accelerated by its dilution in a lighter one or, in the opposite case, a lighter species is decelerated if diluted in a heavier gas. This effect is often defined as aerodynamical acceleration and the dilution process is often called seeding. The difference between the velocity of the carrier gas and the seeded gas is called velocity slip.

The thermodynamic variables of a supersonic beam can be expressed in terms of the Mach number M, defined as

$$M = \frac{v}{v_s}, \tag{2.19}$$

where v is the speed of the beam and $v_s = \sqrt{\gamma k T/m}$ is the speed of sound. One can then write:

$$\frac{T}{T_0} = \left(1 + \frac{\gamma - 1}{2}M^2\right)^{-1}, \tag{2.20}$$

$$v = \sqrt{\gamma \frac{kT_0}{m}} M \left(1 + \frac{\gamma-1}{2}M^2\right)^{-1/2}. \qquad (2.21)$$

Useful expressions can also be written for the pressure and the density before and after the expansion, assuming γ constant,

$$\frac{P}{P_0} = \left(\frac{T}{T_0}\right)^{\gamma/(\gamma-1)} = \left(1 + \frac{\gamma-1}{2}M^2\right)^{-\gamma/(\gamma-1)}, \qquad (2.22)$$

$$\frac{\rho}{\rho_0} = \frac{n}{n_0} = \left(\frac{T}{T_0}\right)^{1/(\gamma-1)} = \left(1 + \frac{\gamma-1}{2}M^2\right)^{-1/(\gamma-1)}. \qquad (2.23)$$

These equations describe the beam properties if M is known and, in particular, if its dependence as a function of the distance from the nozzle is known. This may be calculated from the basic equations of fluid mechanics, i.e., the laws of conservation of energy, momentum and mass combined with the state equation of gases and the first law of thermodynamics (thermal equation state) [2.10].

As mentioned above, the regimes are quite different in the region immediately in front of the nozzle (the so-called sonic region), where the Mach number $M = 1$, and outside the nozzle. Let us first discuss the subsonic and sonic regions. The shape of the nozzle can critically affect the solutions of the mentioned equations because the converging shape can give rise to viscous effects generating the so-called boundary layer. In the regime of fast flow rates and short nozzles, often used to generate highly supersonic beams, these effects are quite negligible, so that the subsonic region of the expansion can be approximated to a quasi one-dimensional compressible flow with M constant across any nozzle cross-section. Under these assumptions, and for an isentropic expansion, at each cross-sectional area A of the nozzle the following relationship holds:

$$\frac{A}{A^*} = \frac{1}{M}\left[\frac{2}{\gamma+1}\left(1+\frac{\gamma-1}{2}M^2\right)\right]^{(\gamma+1)/2(\gamma-1)}, \qquad (2.24)$$

where A^* is the area of the cross-section of the nozzle at the "throat", where the diameter is the minimum. The boundary layer effects cannot be neglected if a diverging section is added to the nozzle since the assumption of an isentropic expansion will break down. Nonetheless, (2.24) still holds up to the nozzle "throat" and gives the required dependence of the Mach number M as a function of the distance from the nozzle, once A is properly expressed. The mass flow rate from a nozzle source can then be written in the form (at the nozzle exit $M = 1$)

$$\dot{m} = \rho v A = P_0 A^* \left\{\frac{m}{kT_0}\gamma \left[\frac{2}{\gamma+1}\right]^{\frac{(\gamma+1)}{(\gamma-1)}}\right\}^{\frac{1}{2}}. \qquad (2.25)$$

Outside the nozzle, the expansion cannot be treated as quasi-one-dimensional and numerical methods should be used to calculate the functional form of the Mach number from the position. Analytical expressions can be obtained by a fitting procedure to the numerical results that can be obtained via different methods, such as the so-called method of characteristics (MOC) [2.12]. Assuming that the beam, far from the expansion, propagates along straight lines radiating in a spherical expansion, as from a point source centered at x_0/d [2.13, 2.14], the Mach number can be written as

$$M = A \left[\frac{x-x_0}{d} \right]^{\gamma-1} - \frac{\frac{1}{2} \left[\frac{\gamma+1}{\gamma-1} \right]}{A \left[\frac{x-x_0}{d} \right]^{(\gamma-1)}}. \tag{2.26}$$

The predictions of this formula are valid only for x/d not too small since near the nozzle the MOC approximation fails. The density of the beam falls off going out of the centerline of the beam approximately as $\cos^4(\theta)$.

The dependence of the Mach number on the position, together with eqns. (2.23) and (2.20) gives, for an isentropic expansion, the evolution of the thermodynamic variables along the beam. Figure 2.4 shows the typical evolution of velocity, temperature and collision rate, normalized to the corresponding values in the stagnation chamber, as a function of the distance from the nozzle along the beam axis x/d.

It should be emphasized that while the velocity reaches the limiting value in a very few nozzle diameters (for $\delta = x/d = 5$, $V = 0.98 V_\infty$) the other

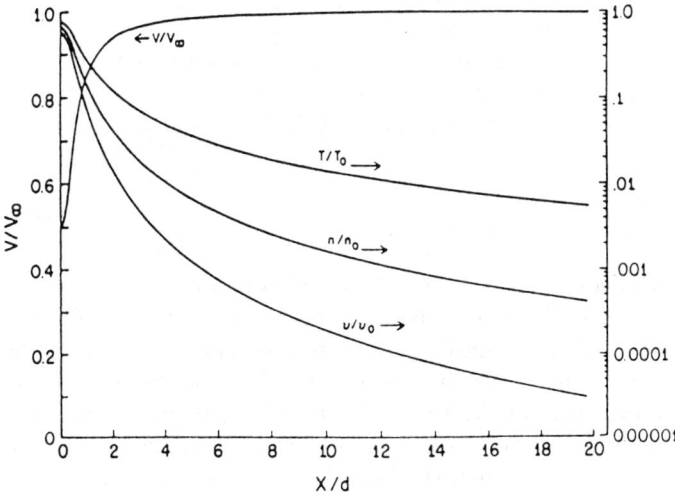

Fig. 2.4. Typical evolution ($\gamma = 5/3$), of velocity V normalized to the limit velocity V_∞; temperature T, and density n normalized to the corresponding values in the stagnation chamber T_0 and n_0 and of the collision rate ν for binary hard spheres. The abscissa reports the distance from the nozzle along the beam axis. (From [2.10])

parameters (temperature, density and collision rate) keep falling off with distance up to 20 and more nozzle diameters.

The T determines the spread in velocity around the mean, so that the resulting energy resolution of the beam becomes proportional to T/V^2. It therefore depends on the position at which the change from continuous to collisional free molecular flow occurs (freezing region after which the beams characteristics no longer change). On the basis of this treatment, the key parameter to compare different nozzle geometries is the mass flow rate. The nozzle giving rise to the same $\dot{m}$ would produce beams with comparable final values of the thermodynamic variables. In the case of channeled nozzles, one would need higher stagnation pressures to reach the same mass flow rate and therefore the same beam characteristics.

Some properties of free-jet expansions arise from nonequilibrium processes. They are related to the drop in collision rate that, via the transition to the free molecular flow, freezes kinetic processes. Examples are the cooling of the internal degrees of freedom and the translational relaxation. In both cases the high collision rate at the beginning of the expansion allows the corresponding kinetic processes to take place, but, towards the end of the expansion, the sudden drop of the collisions and the onset of the free molecular regime produce non-Boltzmann distributions. Other deviations from the ideal isentropic predictions arise from collisions with the beam molecules backscattered by the walls or by the skimmer. They attenuate the beam intensity and rethermalize the velocity distribution producing a heating of the beam in terms of the effective temperature.

In order to obtain the velocity spreads (or temperature) reached before the expansion runs out of collisions, one should model the kinetics. The non-continuum regime can be quantitatively studied by solving the Boltzmann equation, which is not at all a trivial procedure. The solution is very frequently based on a model velocity distribution of the form (ellipsoidal drifting Maxwellian model)

$$f(\bm{v}) = n\sqrt{\frac{m}{2\pi k T_\|}}\left(\frac{m}{2\pi k T_\perp}\right)\exp\left(-\frac{m}{2kT_\|}(v_\| - V)^2 - \frac{m}{2kT_\perp}v_\perp^2\right), \quad (2.27)$$

where the four parameters n, V, $T_\|$, and $T_\perp$ can be obtained by solving the moment equations derived from the original Boltzmann equation [2.10]. The distribution function (2.27) assumes Gaussian profiles ($f(v_\|)$ and $f(v_\perp)$) in two directions about the mean velocity. Deviations have been experimentally demonstrated, however this model form is quite useful to understand the basic features of a supersonic beam and to predict the parallel component determining the energy resolution of the beam. From this point of view even though the Mach number is still a very useful parameter, it is worth introducing the speed ratio S defined as the ratio between the mean velocity V and the velocity thermal spread

$$S = \frac{V}{\sqrt{\frac{2kT}{m}}} = M\sqrt{\frac{\gamma}{2}}. \tag{2.28}$$

Very high speed ratios would provide quasi-monochromatic beams. This can be achieved by pushing the supersonic cooling process to the limit, by increasing the source pressure. A strong limit on achieving this goal is due to the onset of clustering that generates heating of the beam. From this point of view a very special case is He. It does not condense, at typical free-jet expansion conditions, because its dimer is very weakly bound, so that the nucleation process has to go through the direct formation of higher clusters that, as discussed in the following section, is far less probable. In a He beam source one can, therefore, increase the pressure to very high values producing a very strong expansion and high cooling rates so that speed ratios $S > 150$ can be achieved. This corresponds to velocity spreads $\Delta V/V < 1\%$ that make He beams ideally suitable, as a monochromatic source, to carry out diffraction experiments in surface physics.

Calculated and measured speed ratios agree quite well, so that reliable formulas are available in the literature [2.15, 2.16] to calculate S from the source parameters and the physical properties of the gases. In particular, Beijering and Verster [2.16] suggest the following relationship that easily can be extended to polyatomic species

$$S_{\|,\infty} = a\left[\sqrt{2}dn_0\left(\frac{53C_6}{kT_0}\right)^{1/3}\right]^b. \tag{2.29}$$

The parameters a and b can be obtained from a theoretical model or fitted to experimental data: examples are given in Table 2.1 for gases with different γ values. C_6 is the parameter of a Lennard–Jones interaction potential. The whole factor $(53C_6/kT_0)^{1/3}$ may be replaced by a hard-sphere cross-section $(\pi\sigma^2)$, if needed.

Table 2.1. Parameters for calculating final speed ratios [2.16]

γ	a Theory	Exp.	b Theory	Exp.
5/3	0.527	0.778	0.545	0.495
7/5	0.783		0.353	
9/7	1.022		0.261	

Up to now, we have discussed the free-jet expansion and its properties as essentially coming from an isentropic expansion that gives fairly good predictions of the experimental observations. As soon as a skimmer is introduced, severe complications may arise because of its interference with the expansion itself. This is mostly due to backscattering, from the skimmer itself or its supporting wall, of atoms or molecules coming from the expansion.

The intensity (molecules sr^{-1} s^{-1}) predicted by an ideal case where these interferences may be neglected is given by [2.16]

$$I_0 = \frac{\kappa \dot{N}}{\pi}, \qquad (2.30)$$

where the molecule flow rate $\dot{N}$, which can be derived directly from (2.25) for the mass flow rate, at the nozzle is

$$\dot{N} = n_0 F(\gamma) \left(\frac{\pi d^2}{4}\right) \sqrt{\frac{2kT_0}{m}}, \qquad (2.31)$$

with

$$F(\gamma) = \left(\frac{\gamma}{\gamma+1}\right)^{1/2} \left(\frac{2}{\gamma+1}\right)^{1/(\gamma-1)}. \qquad (2.32)$$

These equations are valid even far from the nozzle in the noncontinuum expansion provided that the mentioned interferences from the background and the skimmer are small.

The parameter κ is the peaking factor, that is 1 for an ideal effusive source and can reach a maximum value of a factor 2 larger for $\gamma = 5/3$ ($\kappa = 1.38$ for $\gamma = 7/5$ and $\kappa = 1.11$ for $\gamma = 9/7$). The very high intensities produced by supersonic sources are not due to the peaking factor but rather to the much larger values of $\dot{N}$ compared to effusive sources.

The skimmer perturbations depend on several factors such as its shape, dimension and from the properties of the wall that holds it in position. A useful formula to estimate the fraction of the ideal intensity beyond the skimmer that one can expect is given by Anderson [2.14]

$$\frac{I}{I_0} \cong 1 - \exp\left(-\left\{S(\frac{r_s}{x_q})\left[\frac{x_d}{x_d - x_s}\right]\right\}^2\right), \qquad (2.33)$$

where x_q is the position of the "quitting surface" defined as the surface that separates the two regimes of the expansion: the continuum isotropic regime and the free molecular collisionless regime. The speed ratio $S > 5$ is taken at the quitting surface (terminal speed ratio) while r is the radius of the skimmer opening positioned at $x_s > x_q$ and x_d is the position of the detector. This expression neglects any attenuation produced by the background gas that may be easily introduced by the Beer's law

$$\frac{I}{I_0} = \exp\left(-n\sigma_{\text{tot}} l\right), \qquad (2.34)$$

where σ_{tot} is the total collision cross section between the beam and the background gases (that should be accurately estimated) while l is the length traveled by the beam.

Kinetic processes occurring in atomic or molecular collisions will also take place during the expansion process with the important difference that they will "freeze" at the transition to molecular collisionless flow. Every kinetic process that requires a number of binary collisions to approach equilibrium less than those experienced in a free-jet expansion, will be subject to relaxation effects and then will "freeze" in the molecular beam. The typical number of collisions in free-jet expansion is $\approx 10^3$, so that rotational and vibrational modes of polyatomic molecule, that typically requires 10 to 100 collisions to relax, will be able to exchange energy during the expansion and will cool, up to the point when the process will stop because of the absence of collisions. Vibrational modes of diatomic or small molecules usually require many more collisions (in excess of 10^3) to relax so that there will be no significant vibrational cooling in these expansions. Other kinetic processes, such as chemical reactions, will undergo similar events, although they have been much less studied. Up to now only the rotational cooling of simple molecules has been systematically investigated and further experimental work is needed to fully characterize larger systems including clusters. Figure 2.5 reports the calculations [2.10], based on the method of characteristics, of the expected cooling of the internal temperature of a linear diatomic molecule without any buffer gas. In this case, there is a significant exchange between the internal energy and the external kinetic energy. In the model, this is taken into account by introducing an effective γ that decreases during the expansion from an initial value, called "equilibrium", and a final one, called "frozen". Two cases are shown: the first (continuous line) is the rotational relaxation where

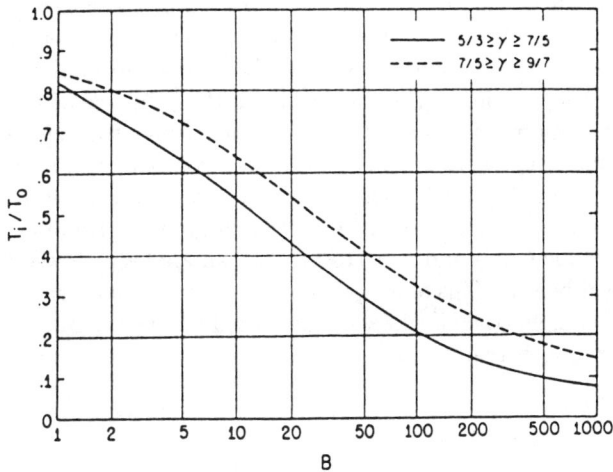

Fig. 2.5. Internal temperature as a function of the dimensionless collision parameter for a diatomic molecule expanding in a supersonic beam without a buffer gas. The continuous line refers to the rotational relaxation while the dotted line refers to the relaxation of a vibrational mode. (From [2.10])

$7/5 \leq \gamma \leq 5/3$, and the second concerns the relaxation of a vibrational mode of the same molecule where $9/7 \leq \gamma \leq 7/5$. The temperature reduction over the entire range of the dimensionless collision parameter B [2.17] is quite consistent and is about a factor 10 of the original temperature in the source.

Another effect that is worth mentioning here is the spatial separation of heavy and light species that will be produced by the supersonic expansion of a mixture. The composition of the beam striking the skimmer is very probably the same as the gas mixture coming out of the source. After the skimmer and during the flight, the centerline flux will favor the heavier species while, moving outward from the central direction of the beam, the density of lighter species increases. This is the consequence of two major effects. Assuming that at the entrance of the skimmer ("quitting surface") a near continuum flow exists, the different species present will have about the same mean velocity and temperature. After the skimmer, in the free molecular flow regime, the perpendicular component of the flight depends on the perpendicular speed ratio given by

$$S_\perp = \frac{V}{\sqrt{\frac{2kT_\perp}{m}}}. \tag{2.35}$$

Consider then what happens to the lighter species: for a given V and $T_\perp$: the lighter species will have a smaller $S_\perp$ corresponding to a larger component of perpendicular motion producing a larger lateral spread. As a consequence the centerline density of heavier species will be larger. The other effect is due to the fact that the skimmer walls will block light species coming from different points of the quitting surface more efficiently [2.10, 2.18].

Significant separation effects can be obtained with seeded beams containing large particles. These effects were systematically studied by Becker et al. [2.19]. To explain these observations, Reis and Fenn [2.20] proposed a mechanism by which the separation is a consequence of the presence of the probe (skimmer) on the beam trajectory. The deceleration and the severe streamline curvature caused by the bow shock detached from the skimmer, favor the separation of the light particles which are diverted from the center of the beam, whereas the heavy ones can follow their original trajectories. The role of particle inertia led Reis and Fenn to propose an analogy between seeded beams and aerosol beams. This analogy seems to be very appropriate: aerosol impactors exploiting inertial effects are widely used for the separation of particles with narrow size distributions [2.21, 2.22].

2.1.3 Pulsed Beams

Pulsed molecular beams derive their unique features from the fact that the source is normally off and it is turned on only for a brief interval. The major reason to adopt pulsed molecular beams is the achievable peak intensity which is considerably higher compared to a continuous source operated at the same conditions (pumping speed Z, source chamber volume Ω, nozzle source diameter d_p and d_{cw} for the pulsed and the continuous sources respectively). Let us assume that both sources work in a supersonic flow regime and that their volume flow rates depend only on the nozzle cross-sectional areas and not on the respective source pressures P_{0p} and P_{cw}. One can also suppose that their final mean velocities (for sufficiently high Mach numbers) are equal and that no significant differences would be present in the angular distributions of the two beams. Then the ratio of the respective instantaneous intensities I or densities n are given by

$$\frac{n_p}{n_{cw}} = \frac{I_p}{I_{cw}} = \frac{Q_p P_{0p}}{Q_{cw} P_{cw}} = \frac{P_{0p} d_p^2}{P_{0cw} d_c^2}, \quad (2.36)$$

where Q_p and Q_{cw} are the volume flow rates for the pulsed and the continuous source respectively. For a continuous source the background pressure due to the beam is obtained by equating the flow rate coming from the nozzle and the gas load on the pump:

$$P_{bcw} = P_{0cw}\left(\frac{Q_{cw}}{Z}\right). \quad (2.37)$$

For a pulsed beam, we introduce the duty factor ξ, defined as the ratio of the pulse duration Δt and the pulse period β (square uniform pulse approximation)

$$\xi = \frac{\Delta t}{\beta} \quad (2.38)$$

that will produce a time-dependent background pressure the average of which can be written as

$$P_{bp} = \xi P_{0p}\left(\frac{Q_p}{Z}\right). \quad (2.39)$$

The best way to compare the performance of the two kinds of beams is to consider the ratio between the intensity and the background for each of them. This is in fact the meaningful parameter for beam experiments. In particular, for scattering and spectroscopic measurements it becomes the key factor to determine the signal to noise ratio. For a pulse source this ratio $\langle S_p \rangle$ will be obtained by considering the peak-pulse intensity and the average background pressure while for the continuous beam S_{cw} the steady state values should be taken into account:

$$\frac{\langle S_p \rangle}{S_{cw}} = \frac{\frac{I_p}{\langle P_{bp} \rangle}}{\frac{I_{cw}}{P_{bcw}}} = \frac{1}{\xi}. \quad (2.40)$$

To give an idea of how large this ratio could be, let us consider a pulsed beam with pulses of 100 μs and a repetition rate of 10 Hz, then one has $\langle S_p \rangle / S_{cw} = 10^3$.

In typical beam experiments a simple rule should be kept in mind: *whenever one element in an experiment is pulsed, the best signal to noise ratio will* be achieved by pulsing all elements [2.23]. If this is done, instead of considering the average background pressure, one should consider the time-dependent values which are a function of the pump-out time of the vacuum system. Through a simple calculation one finds the following result:

$$\frac{S_\mathrm{p}}{S_\mathrm{cw}} = \xi^{-1} \left(\beta \sum_{n=1}^{\infty} \mathrm{e}^{-n\beta/\tau} \right)^{-1} \qquad (2.41)$$

where τ is the pump-out time constant for the source chamber [2.23].

The factor in parentheses takes into account the summing of the tail of one pulse with the rise of the next pulse: in the limit of $\beta/\tau \longrightarrow 0$ (very high source repetition rates or very slow pump-out times) the pulses will sum each other giving rise to a background pressure which is almost constant and equivalent to that of (2.39). The other regime, where each pulse is actually pumped away before the next one is produced, is the most interesting one. In this regime the ratio $S_\mathrm{p}/S_\mathrm{cw}$ becomes much larger then ξ^{-1}. Assuming again a source producing pulses of 100 μs at a repetition rate of 10 Hz, if the pump-out time is of the order of 10 ms, the enhancement factor becomes 2×10^6.

Another interesting characteristic of pulsed expansions arises from the position of the Mach disk given by (2.14). As we have already seen, for a continuous beam the Mach disk is quite close to the nozzle. Due to the low background pressure achievable with pulsed sources, the Mach disk is very far from the nozzle and very often, for typical experimental conditions, it is at distances larger then the vacuum chamber dimensions. Provided that the average background is kept low enough (this can be difficult to achieve when using long pulses), the direct effects due to the shock wave would therefore not be present. Indirect effects due to reflections from the chamber walls, would not produce any effect since the beam pulse would have already passed the skimmer before reaching the beam path.

2.2 Nucleation and Aggregation Processes

2.2.1 Classical Theory

Studies of nucleation, the process through which clusters of a stable phase grow in a metastable phase, have been the object of continuous interest with the aim of developing an accurate theory that would interpret the early stages of phase transformations, such as condensation of gases or solidification of melts. These theoretical efforts have given rise to the so-called classical condensation theory based on the original work of Volmer and Weber [2.24]. A series of contributions of several authors, including Zeldovich [2.25], Frenkel [2.26], and Turnbull [2.27, 2.28], show that the classical theory is capable of predicting the critical supersaturation of gases, that is the degree of supersaturation at which liquid droplets form.

The classical condensation theory is still very useful since it offers the advantage of including very few parameters, derived from basic concepts.

Tests of the ability of this model to predict nucleation rates, that can now be measured with high accuracy [2.29], have shown the limits of this approach: even though the nucleation rate as a function of supersaturation is predicted with a reasonable accuracy, the temperature dependence gives nucleation rates that are underestimated at low temperatures and overestimated at high temperatures [2.30].

The state of the art of the theoretical approaches is discussed in detail by Wu [2.31], while classical theory is discussed in more detail by Christian [2.32] (homogeneous nucleation) and by Sigsbee [2.33] (heterogeneous nucleation), here, we only summarize the basic concepts and results in order to give to the reader a framework. The following discussion will be focused in particular to the aspects more strictly related on cluster beam formation.

2.2.2 Homogeneous Nucleation by Monomer Addition

We will restrict the following discussion to the case of gas to liquid transitions, which is the one most relevant for cluster sources. On the other hand, since the only clusters that can be used efficiently for producing beams are those forming in the vapor, we will describe only homogeneous nucleation.

Classical theory, in this case, is based on the fundamental assumption that, independently of the number of atoms n that it may contain, the properties of a cluster (assumed to be a liquid) can be derived by extrapolating those of the corresponding bulk liquid. On this basis, the calculation of the minimum work of formation of a cluster from a metastable phase yields the partition of the free energy. This can be written as the sum of two terms, one bulk and the other surface, both having energy densities constant and equal to the corresponding macroscopic values. This is the so-called capillary approximation giving

$$G(R) = 4\pi R^2 \gamma + \frac{4}{3}\pi R^3 \left(\frac{\mu_l - \mu_g}{V_c}\right), \qquad (2.42)$$

where G is the Gibbs free energy and R the radius of a cluster containing n monomers. The first term on the right side of (2.42) is the surface term where γ is the (macroscopic) surface energy per unit area at the gas-liquid interface. In the second, the bulk term, μ_l and μ_g are respectively the chemical potential of the two phases involved (liquid and gas) while V_c is the volume per monomer in the condensed bulk phase. The cluster, under the assumptions made, can be assumed spherical so that the relationship between R and n is

$$R = \left(\frac{3}{4\pi} V_c n\right)^{\frac{1}{3}}. \qquad (2.43)$$

We can now rewrite (2.42) as a function of the cluster size n:

$$G(n) = \sigma n^{2/3} - \delta n, \tag{2.44}$$

where $\sigma = (36\pi V_c^2)^{1/3}\gamma$ and $\delta = \mu_g - \mu_l$ assume positive values since the liquid is stable.

We can gain an idea of the typical behavior of G as a function of the cluster size by analyzing the two terms σ and δ. For an ideal supersaturated gas, σ is positive by definition and δ is given by

$$\delta = kT \ln \zeta, \tag{2.45}$$

where $\zeta = P/P_e$ is the supersaturation ratio (or degree of supersaturation) that is the ratio between the actual pressure P and the equilibrium pressure P_e at the temperature of interest. Since also δ is positive, the relationship (2.44) should have a maximum as shown schematically in Fig. 2.6 where G is plotted versus n for a typical pair of values of the parameters σ and δ. The size n^* corresponding to the maximum value of the free energy G^* is a critical size that separates the region $n < n^*$ of the clusters (called embryos) that increase their free energy while growing from the range of $n > n^*$ where the clusters (called nuclei in this case) will decrease their free energy while growing. Nucleation is therefore a barrier-crossing process where the critical mean free energy $G(n^*) = G^*$ represents the barrier that an embryo must overcome to grow to an observable size.

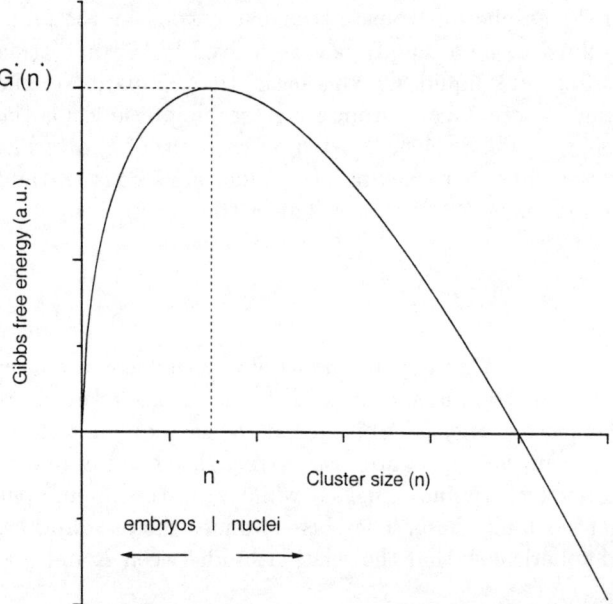

Fig. 2.6. Gibbs free energy as a function of the cluster size n. The size n^*, corresponding to the maximum G^*, is the so-called critical size that separates embryos from nuclei of condensation

2.2 Nucleation and Aggregation Processes

The processes responsible for the formation of clusters are assumed to be bimolecular reactions (such as monomer addition) and evaporation, that compete to give the final cluster distribution. If we assume that the density of monomers dominates, so that the collision rates with dimers, trimers and larger clusters are not high, we can adopt the monomer approximation so that the basic scheme is

$$A + A_n \underset{\alpha_n}{\overset{\beta_n}{\rightleftarrows}} A_{n+1} \qquad (2.46)$$

where A is the monomer, A_n is the cluster of size n. β_n and α_n are the reaction rates for cluster growth (via monomer addition to the cluster of size n) and for cluster shrinking (via monomer evaporation), respectively.

Since, in classical theory, the sticking coefficient monomer–cluster of size n is assumed to be 1, the cluster growth rate β_n reduces to the number of collisions per unit time that A_n undergoes with the monomers. In the hard sphere framework, this can be easily derived from kinetic theory in the form

$$\beta_n = \sqrt{\frac{kT}{2\pi m}} f_1 (36\pi V_c^2)^{1/3} n^{2/3} (1 + n^{-1/3})^2 (1 + n^{-1})^{1/2} , \qquad (2.47)$$

which is obtained assuming the cluster to be a sphere and including the effects of its translational motion [taken into account by the factor $(1 + n^{-1})^{1/2}$], f_1 being the volume density of monomers.

As far as the evaporation rate α_n is concerned, there is no simple general formula available so that a further assumption (the so-called *constrained-equilibrium*) is made in classical theory. It is based on the idea that a metastable gas phase obeys the laws of fluctuation thermodynamics.

The net growth rate by monomer addition can be expressed in terms of the cluster distribution function f_n (that is the volume density of n-mers) and of the growth and shrinking rates β_n and α_{n+1} in the form

$$J_{n+1/2} = \beta_n f_n - \alpha_{n+1} f_{n+1} . \qquad (2.48)$$

Assuming equilibrium with the bulk liquid (saturated gas phase), the cluster distribution function should be the equilibrium one, satisfying detailed balance so that all the net rates (fluxes) of (2.48) should be zero. The equilibrium distribution function z_n would satisfy then the following relationship:

$$0 = \beta_n z_n - \alpha_{n+1} z_{n+1} . \qquad (2.49)$$

The constrained equilibrium hypothesis assumes that this zero-flux distribution can be still defined for a supersaturated gas on the basis of (2.49). In this way a mathematical definition of the evaporation rate could be obtained independently of the specific form of z_n:

$$\alpha_{n+1} = \beta_n \frac{z_n}{z_{n+1}} . \qquad (2.50)$$

Fluctuation thermodynamics gives then the following form for the zero-flux distribution:

$$z_n = f_1 e^{-G(n)/kT}, \qquad (2.51)$$

where the exponential factor $e^{-G(n)/kT}$ gives the relative probability of fluctuations characterized by a change in free energy $G(n)$; k being the Boltzmann constant.

The net rate or flux (2.48) can now be rewritten by the appropriate substitutions as

$$J_{n+1/2} = -\beta_n z_n \Delta\left(\frac{f_{n+1/2}}{z_{n+1/2}}\right), \qquad (2.52)$$

where the operator Δ is the so called one-step central finite difference operator that, for an arbitrary couple x,y, is defined by

$$\Delta x_y = x_{y+1/2} - x_{y-1/2}. \qquad (2.53)$$

The evolution of the cluster distribution function is obtained from a time dependent theory with appropriate boundary conditions. Because of the continuity principle, the variation in the cluster distribution is given by

$$\frac{df_n}{dt} = -\Delta J_n = \Delta\left(\beta_{n-1/2} z_{n-1/2} \Delta\left(\frac{f_n}{z_n}\right)\right) \qquad (2.54)$$

which is a non-linear rate equation because both β_n and z_n are functions of the monomer concentration f_1 that is of course varying in time. In the limit of an infinite supply of monomers (or else in the case of a condensation process that does not perturb too much the original vapor), one may assume that the original population of monomers does not vary significantly. This is the simplest approximation used in classical theory:

$$\frac{df_1}{dt} \approx 0 \iff f_1(t) = f_1(0). \qquad (2.55)$$

Equation (2.54) would describe then the evolution of the cluster distribution reaching, for a constant temperature T, a steady state. For $n \geq 2$, $df_n/dt = 0 = -\Delta J_n$:

$$\Delta J_n = 0 = J_{n+1/2} - J_{n-1/2} \iff J^{ss} = J_{n+1/2} = \text{const}. \qquad (2.56)$$

This equation defines the steady state flux J^{ss}. Of course this describes an idealized picture in which the population of monomers is infinity. Otherwise there would not be a real steady state because, after a while, the condensation process will produce a depletion in the monomer population that in turn will bring back the system to an equilibrium where no supersaturation is present. The approximation used in this classical approach is therefore describing an isothermal state that can be considered stationary only in the approximation that both fast, transient like events, and slow processes, such as the depletion of population of the monomers, can be discarded or are unimportant. The steady state described by the classical theory, would therefore be limited to an intermediate time-scale where these hypothesis could be considered valid.

In order to calculate J^{ss} another assumption should be made regarding the largest cluster that can be formed. The idea is that very large cluster $A_{n_{\max}}$ over a certain size $n_{\max}$ have a very small probability of formation and therefore can be discarded as soon as they are formed, so that

$$f_{n_{\max}}(t) = 0 \,. \tag{2.57}$$

We can now rewrite the solution for the flux at steady state in a more manageable form by dividing (2.52) by $\beta_n z_n$ and summing over the cluster sizes, taking into account the boundary conditions (2.55) and (2.57):

$$J^{ss} = \left(\sum_{n=1}^{n_{\max}-1} \frac{1}{\beta_n z_n} \right)^{-1} . \tag{2.58}$$

The steady state cluster size distribution can then be written as a function of the rate J^{ss} in the form

$$f_n^{ss} = z_n J^{ss} \sum_{p=n}^{n_{\max}-1} \frac{1}{\beta_p z_p} \,. \tag{2.59}$$

As mentioned earlier a critical cluster size n^* divides embryos from nuclei clusters so that one should define the *nucleation rate* as the flux $J_{n*+1/2}$ (critical flux). More often the smallest detectable cluster $A_{n_{\det}}$ is defined, the production rate of which is given by the flux $J_{n_{\det}-1/2}$ that would be equal to the critical flux at the steady state.

The calculation of (2.58) for the rate J^{ss} can be further simplified by converting the sum in an integral and integrating it with the help of the so-called steepest descent method [2.31]. The classical nucleation theory gives then an approximate but useful final formula for J^{ss} that is valid for both gas and condensed phases:

$$J^{ss} \approx f_1 \beta^* e^{-G^*/kT} \sqrt{\frac{-1}{2\pi kT} \frac{d^2 G^*}{dn^2}} \tag{2.60}$$

In the case of an ideal gas, also taking into account the expression for the growth rate β given by (2.47), one can easily find a simpler expression:

$$J^{ss} = f_1^2 (1 + n^{*-1/3})^2 (1 + n^{*-1})^{1/2} e^{-G^*/kT} \sqrt{\frac{2V_c^2 \gamma}{\pi m}} \,, \tag{2.61}$$

where the critical parameters assume the following forms:

$$n^* = \frac{32}{3} \frac{\pi V_c^2 \gamma^3}{(kT \ln \zeta)^3} \,, \tag{2.62}$$

$$G^* = \frac{16}{3} \frac{\pi V_c^2 \gamma^3}{(kT \ln \zeta)^2} \,.$$

2.2.3 Homogeneous Nucleation by Aggregation

The classical model, as described up to now, is based on the growth mechanism of monomer addition which assumes a very small density of clusters and an infinite supply of monomers. A generalization of the theory to more complicated growth processes, where agglomeration between clusters is allowed, is, however, straightforward with some adjustments on the assumptions made.

The cluster of size n can now undergo a series of processes $i = 1, 2, \ldots, g$ corresponding to the agglomeration (reaction rate β_n^i) or evaporation (reaction rate α_n^i) of an aggregate of size i:

$$A_n + A_i \underset{\alpha_n^i}{\overset{\beta_n^i}{\rightleftarrows}} A_{n+i}, \tag{2.63}$$

where g is the largest size of the partner cluster that will coalesce. The nth rate of growth, of the cluster of size n, could now be calculated, in analogy to (2.48), by summing on all the partner clusters that could coalesce or be evaporated to produce the n-mer:

$$J_{n+1/2} = \sum_{l=1}^{g} \sum_{j=n-l+1}^{n} (\beta_j^l f_j - \alpha_j^l f_{j+l}). \tag{2.64}$$

Following a procedure analogous to that described for the monomer addition growth we can write approximate formulas to calculate β_m^l and α_m^l. Using again kinetic theory for hard spheres, the growth rates assume the following form:

$$\beta_j^l = \sqrt{\frac{k_B T}{2\pi m}} f_l (36\pi V_c^2)^{1/3} j^{2/3} \left[1 + \left(\frac{l}{j}\right)^{1/3}\right]^2 \left(1 + \frac{l}{j}\right)^{1/2}. \tag{2.65}$$

The constrained equilibrium hypothesis is the framework of approximations that allows us to obtain a simple formula also for cluster evaporation rate coefficients, as already discussed for the growth by monomer addition:

$$\alpha_j^l = \beta_{j-l}^l \frac{z_{j-l}}{z_j}. \tag{2.66}$$

The capability, in the framework of the classical theory, of handling only time independent rate coefficients, gives rise to quite severe boundary conditions where all the (l) cluster distributions, up to $l = g$ are kept constant. The concentration of monomers is again assumed to be constant ($f_1(t) = f_1(0)$) while all the others up to $l = g$ are left to equilibrate rapidly with the monomers so that the constrained values will then be

$$f_l(t) = f_1(0) e^{-G(l)/kT} \iff l = 1, 2, \ldots, g. \tag{2.67}$$

This approach, even though it appears to be not very realistic, offers the great advantage that one can find steady state solutions based on matrix methods that can be easily calculated. As shown for monomer addition, we can find the steady state solutions by introducing boundary conditions defining a maximum size of the formed clusters (see (2.57)) and applying the continuity equation.

We can define a vector of cluster concentrations

$$\mathbf{f} = \begin{pmatrix} f_{g+1} \\ f_{g+2} \\ . \\ . \\ . \\ f_{n_{max}-1} \end{pmatrix} \tag{2.68}$$

and its time evolution would follow a matrix equation of the form

$$\frac{d\mathbf{f}}{dt} = -B\mathbf{f} + \mathbf{f}^0 \tag{2.69}$$

where the matrix operator B is a band matrix with $2g+1$ bands and the constant vector $\mathbf{f}^0$ gives the contribution to the flux coming from the distributions of the sizes $n = 1,..,g$ that are time independent (see (2.67)) The steady state solutions can now be written in a matrix form

$$\mathbf{f}^{ss} = B^{-1}\mathbf{f}^0 \tag{2.70}$$

that, in turn, gives the steady state nucleation rates.

2.2.4 Nucleation of Clusters in Beams

In a molecular beam a straightforward application of the classical nucleation theory will not give a good description of the clustering processes, even for the simplest case of a gas-aggregation source. In fact, the processes involved in the cooling of the vapor into the cell are not easy to model in this framework. In particular it is difficult to define accurately the thermodynamic variables, since the cooling is obtained by mixing hot vapors in a cold buffer gas (see Chap. 3). This process is usually associated with combined laminar and turbulent convection flows that are difficult to predict in space and time and make any modeling very difficult. Nonetheless, since the cooling rate in a gas-aggregation cell is low enough ($dT/dt < 5 \times 10^6\,\mathrm{K\,s^{-1}}$), the steady state nucleation theory is applicable, provided that the definition of the thermodynamic variables is accurate enough for the specific source and operating conditions.

Some assumptions, corresponding to well defined source design and operation, can make the model quite simple with fairly good predictions on the beam characteristics. In particular one may assume:

(i) very small concentrations of vapor atoms and clusters compared to the buffer gas, so that the flow may be considered stationary and unidimensional;

(ii) the existence of a section of the source, close to the nozzle exit, where the thermalization process is complete and where the clustering takes place, characterized by a unique temperature for the vapor and the buffer gas ($T_\mathrm{g} = T_\mathrm{v}$);

(iii) the buffer gas is in thermal equilibrium with the cell walls.

Rechsteiner and Ganière [2.34], assuming for the clusters the same surface tension of the bulk, use the classical homogeneous condensation theory for Li cluster beams produced by a gas-aggregation source with a converging nozzle and a He buffer gas. Even though the hypothesis of the existence of a region of complete thermalization appears to be crude, some basic experimental trends are reproduced and the effects and limits of the approximation made are discussed. The model fails when the expansion into the vacuum is pushed towards a supersonic expansion limit. For rapid changes in density (Knudsen number) and temperature (Mach number), the assumptions inherent in nucleation rate theory (time-independence, isothermicity and local thermodynamic equilibrium) break–down.

A way to overcome these difficulties is to introduce a time-dependent nucleation rate by defining the so-called nucleation lag time τ_n that is a function of the thermodynamic state variables [2.35]. This new time dependent nucleation rate J^t could be written as a function of the steady state one J^ss and of time t in the form

$$J^\mathrm{t} = J^\mathrm{ss} \left[1 - \exp\left(-\frac{t}{\tau_n}\right)\right]. \tag{2.71}$$

Figure 2.7 shows a typical behavior (for a beam of Au seeded in Ar) of J^ss, J^t, and of the corresponding mass fraction condensed (g^ss and g^t) as a function of the distance from the nozzle in units of nozzle diameter $x^* = \delta$, as calculated by Stein [2.36]. During the expansion along the beam axis, due to the sharp increase of the supersaturation ς, J^ss undergoes a very sharp rise followed by a sudden decrease in the span of slightly more than a nozzle diameter. In the case of a Laval nozzle this effect (shutting off the nucleation) is due to the fall in supersaturation while in a free-jet expansion is triggered by the sharp decrease in density. In the framework of this model the correct nucleation rate would then be given by J^t until it crosses J^ss and then by the values of J^ss that becomes the correct one after the crossover. J^t is typically several order of magnitude smaller then J^ss.

As discussed by Stein [2.36] the nucleation approach works well only when the rate of change of state is slow enough to maintain both thermodynamic equilibrium and a quasi-equilibrium cluster distribution up to the critical size for nucleation (where $J = J^\mathrm{ss}$ and there is no nucleation lag–time). This is the case of a very slow expansion typical of gas-aggregation sources with cooling rates typically $\mathrm{d}T/\mathrm{d}t \leq 5 \times 10^6 \, \mathrm{K\,s^{-1}}$.

As the supersonicity of the jet increases and the cooling rate is in the range $5 \times 10^6 < \mathrm{d}T/\mathrm{d}t < 5 \times 10^7 \, \mathrm{K\,s^{-1}}$, then the steady state value J^ss gives only an underestimate of the nucleation rate J even though it is still a reasonable approximation. As soon as the expansion becomes more rapid, with cooling rates in the range $5 \times 10^7 < \mathrm{d}T/\mathrm{d}t < 5 \times 10^8 \, \mathrm{K\,s^{-1}}$ the time-dependent

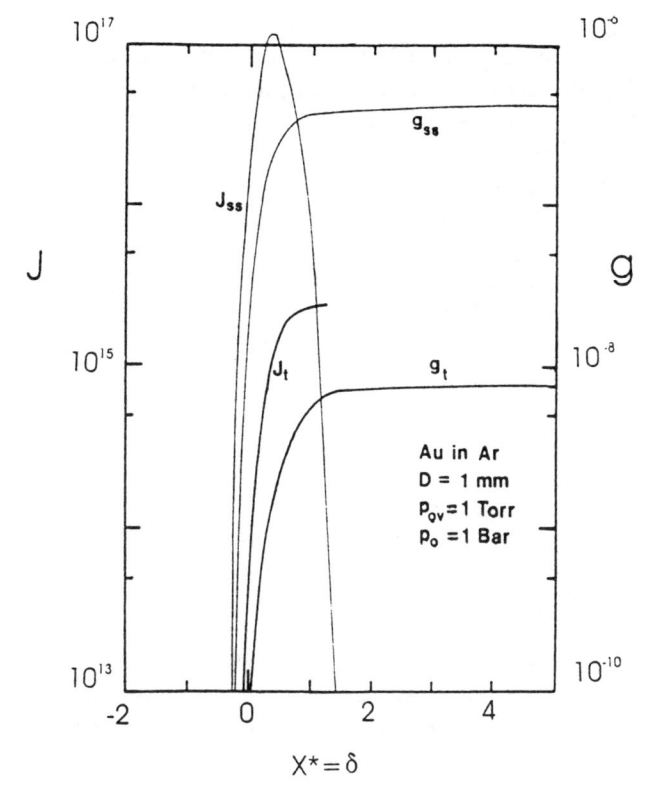

Fig. 2.7. Steady state (J^{ss}) and time dependent (J^t) nucleation rates as a function of the distance from the nozzle in units of nozzle diameters ($x^* = \delta$). The plot shows also the corresponding behavior of the condensed mass fractions (g^{ss} and g^t). The data refer to the case of Au seeded in Ar. (From [2.36])

nucleation theory should be used calculating $J = J^t$ on the basis of the time lag model. However, even this approach begins to fail, giving only an approximate indication on the nucleation process occurring, for nozzle expansions where the cooling rates are one order of magnitude larger. For free-jet nozzles giving rise to extremely high cooling rates ($10^9 < dT/dt < 10^{11}\,\mathrm{K\,s^{-1}}$) not even the time lag approach is valid. Because of the rapid fall off of the density (large Knudsen numbers), the collision rate no longer guarantees the assumption of quasi equilibrium in the cluster distribution and perhaps that is not high enough even to maintain thermodynamic equilibrium.

The time lag approach, coupled with the equation for supersonic compressible gas flow, can be considered adequate for Laval-type contoured nozzles that are characterized by relatively slow expansions. However, some care has to be taken in considering explicitly viscous boundary layer effects.

Rao and Smirnov [2.37] have analyzed in detail the kinematics of clustering in a free-jet expansion taking into account the various processes occurring during the expansion: cluster evaporation and attachment of atoms to clusters; the aggregation of two clusters; three-body collisional processes involving atoms and the thermal processes occurring in the course of vapor expansion. In particular, the kinetics of cluster growth as a result of vapor expansion is analyzed theoretically by coupling the collisional processes involving clusters [2.38, 2.39], with the thermal processes corresponding to both the cooling of the vapor due to its expansion and the heating produced by the clusters formation and growth. The authors assume that the collision of two clusters is similar to that of two liquid drops so that they can calculate the rate constant of monomer attachment and cluster aggregation processes. The complex formed by the collision is a cluster in which the internal energy of the colliding clusters is redistributed on its own degrees of freedom and has a long lifetime so that it can participate to further clustering processes. They show how important are the first stages of nucleation with the formation of molecules that act as nuclei of condensation for the following clustering processes. Since they occur via three-body interactions, this step may be a bottleneck to the cluster growth so that the initial presence of molecules affects strongly the final cluster distribution in the beam.

The information given by this model is restricted to average properties of the beam because of the approximations implicit in the liquid drop model. Despite this, the analysis appears to be of quite general applicability while a strong limit is that its accuracy is based on the knowledge of the physical properties of diatomic and small clusters, which are not often available and accurate enough.

2.2.5 Semi-empirical Approach to Clustering in Free Jets

In a supersonic free jet the initial thermodynamic state of the gas prior to the expansion, together with the size and shape of the nozzle, determine whether the clusters are formed or not during the expansion. In particular, the cluster formation is favored by the increase of the density of the gas n_0 into the source, a decrease of the temperature of the source T_0, and by the use of large nozzles.

As we have seen in the previous sections, a rigorous theory to describe such a process is not available since classical nucleation models are inadequate to describe clustering in free-jet expansions [2.40]. Nonetheless, it is possible to develop scaling laws correlating flow fields that produce the same cluster distribution by accounting for the changes in kinetic conditions for cluster formation as a function of n_0, T_0, and d [2.41]. On the other hand, it is known that the formation of clusters for gases that have the same type of intermolecular potential [2.16, 2.42, 2.43] gives rise to the concept of corresponding free jets. From the development of these aspects one can obtain an insight into clustering in a supersonic expansion with quite realistic predictions of the characteristics of the beam [2.44].

2.2 Nucleation and Aggregation Processes

As stated in Sect. 2.1.2, in a free jet the collisional regime region scales with $n_0 d$: the expansion can be assumed to be isentropic so that the density will drop with the temperature following the relationship

$$\frac{n}{n_0} = \left(\frac{T}{T_0}\right)^{\frac{1}{\gamma-1}}, \tag{2.72}$$

where $\gamma = 5/3$ for a monatomic gas. For the sake of simplicity we restrict the discussion to this case in the following treatment.

It is possible to introduce a scaling parameter Ψ given by

$$\Psi = n_0 d T_0^{1.25}, \tag{2.73}$$

which determines the final temperature of the beam for bimolecular collision processes. Sources characterized by the same Ψ will give rise to the same final beam temperature (T_∞) depending on the collision processes. In the simplest assumption of hard sphere atoms, since the product of the mean free path times the density (λn) becomes independent from temperature, a very simple form can be derived [2.45]

$$T_\infty = \frac{K}{\Psi^{0.8}}, \tag{2.74}$$

where the parameter K is specific for each gas.

Using this scaling relationship it is possible, for a given gas, to define the source conditions that determine the same final beam temperature. This approach can also be generalized to define the source conditions, for a given gas or vapor, leading to the same cluster mass distribution.

To this end, it is assumed that the formation of clusters is a homogeneous gas reaction regulated by bimolecular collisions between atoms and the cluster, while the decay of the cluster goes through evaporation of one atom (unimolecular decay). On this basis, a clustering scaling parameter Γ can be defined correlating source conditions that, for a given gas, yields the same cluster distribution [2.41, 2.45]

$$\Gamma = n_0 d^q T_0^{0.25q-1.5} \iff 0.5 < q \leq 1. \tag{2.75}$$

The upper limit $q = 1$ corresponds to $\Gamma = \Psi$, that is to bimolecular reactions.

This model is semi empirical so that the parameter q cannot be derived from the theory but it should be determined experimentally by measurements with different nozzle diameters d at fixed source temperature T_0. The scaling parameters Γ and Ψ can be used for rare gas atoms and for metal vapors as far as the adiabatic cooling and the clustering process are regulated by binary collisions and unimolecular decay.

A much more difficult task is to correlate the behavior of different gases and vapors. For rare gases a relatively simple concept of corresponding jets has been derived [2.42] and shown to be experimentally confirmed. It is based on the thermodynamic similarities of gases in corresponding states combined with the similarity in the gas-dynamics of expansions with the same Knudsen number. A generalization of these concepts to metals is quite difficult: even among the alkali metals themselves there are not simple similarities to use [2.44]. It is thus necessary to find a set of characteristic values for each substance so that the condensation parameters can be written in a non dimensional form. A choice that has the advantage of using properties that are known for all the substances of interest is the following:

$$r_{ch} = \left(\frac{m}{\rho}\right)^{\frac{1}{3}}, \tag{2.76}$$

$$T_{ch} = \frac{\Delta H_0^0}{k}, \tag{2.77}$$

where m and ρ are the atomic mass and the density in the solid of the substance, while ΔH_0^0 is the sublimation enthalpy per atom at $0\ K$. On the basis of the characteristic units defined by (2.76) and (2.77) one can define a set of dimensionless variables:

$$n_0^* = n_0 r_{ch}^3, \quad d^* = \frac{d}{r_{ch}}, \quad T_0^* = \frac{T}{T_{ch}}, \quad p^* = \frac{p}{p_{ch}} = \frac{p r_{ch}^3}{k T_{ch}}. \tag{2.78}$$

From the definition of the scaling parameters Ψ (2.73) and Γ (2.75) combined with the definition of the variables (2.78), the scaling parameters can be written in a dimensionless form:

$$\Psi^* = \frac{\Psi}{\Psi_{ch}} = \Psi T_{ch}^{1.25} r_{ch}^2, \tag{2.79}$$

$$\Gamma^* = \frac{\Gamma}{\Gamma_{ch}} = \Gamma T_{ch}^{(1.5-0.25q)} r_{ch}^{3-q}. \tag{2.80}$$

For each specific expansion and vapor one can calculate the characteristic units and then the scaling parameters that will provide direct information about the degree of clustering. The calculated values for a series of elements (metals and rare gases) are reported in [2.44].

The comparison with the experimental data gives the value of q to be used, and it also gives a general rule that helps in obtaining information about the clustering that one may expect from a source designed for a specific vapor. In particular, three major ranges of Γ^* define quite different source beam characteristics: for $\Gamma^* < 200$ no clustering is expected, for $200 < \Gamma^* < 1000$ onset of clustering up to the formation of large clusters, for $\Gamma^* > 1000$ massive condensation with cluster sizes larger than 100 atoms/cluster.

The semi-empirical definition of scaling parameters is very useful in designing a source and predicting its behavior. The correlation with the available data is reasonable so that calculations based on this scaling parameter can really give information about the condensation process in free jets even for a very fast expansion. A major limit of this approach is that it gives only a probability for clustering under certain source conditions, on the basis of monomer addition, so that it could not predict the expected maximum cluster size and will not work for materials such as molecular vapors.

Weiel [2.46] proposed a model for seeded beams with highly diluted vapors, where the effects due to aggregation may be important in the growth of larger clusters. Since one looks only for an upper limit to the cluster size achievable at given source conditions and for a given vapor, some simple assumptions can be made: every collision (between monomers, monomers and the cluster and between two clusters) is effective in forming a new cluster, the heat of condensation released by the cluster growth is efficiently carried away by the collisions with the carrier gas (very high dilution). Moreover, as in the semi empirical model of Hagena [2.45], the shape of the nozzle (sonic or conical) can be explicitly taken into account by defining an effective nozzle diameter.

From the calculation of the collision rate in an isentropic expansion and considering a monatomic carrier gas, Weiel [2.46] derives the following relationship for the cluster growth along the axis of the expansion:

$$\frac{dN}{d\delta} = 4dd_1^2(\ln 2)n_1(0)\left(\frac{3\pi m_{\text{gas}}}{5m_1}\right)^{1/2}$$
$$\times \left[1 + \frac{M^2(\delta)}{3}\right]^{-3/2} M^{-1}(\delta)N^{1/6}(\delta) \,. \tag{2.81}$$

where δ is the ratio between the distance from the nozzle and the nozzle diameter d, $d_1 = $ (mass of monomer/density of bulk material)$^{1/3}$ is the diameter of a monomer of the vapor, $n_1(\delta)$ is the number density of the vapor component in the expanding gas if no clustering occurs, m_{gas} is the atomic mass of the carrier gas, m_1 is the atomic or molecular weight of the clustering material, $M(\delta)$ is the local Mach number during the expansion.

In order to evaluate (2.81) one should have an explicit form for the local Mach number that is related to the respective area ratio $F(\delta)/F^*$ of the streamtubes by the following relationship:

$$\frac{F(\delta)}{F^*} = \frac{\left(\frac{(M^2(\delta)+f)}{(1+f)}\right)^{\frac{(1+f)}{2}}}{M(\delta)} \,, \tag{2.82}$$

where f is the number of degrees of freedom of the expanding gas mixture ($f = 3$ for a monatomic carrier gas).

As will be discussed in more detail in Chap. 3, describing source design, a conical shaped nozzle is particularly well suited to produce intense cluster beams. In this case, the respective area ratio (2.82) can be approximated by the area ratio of a hyperbolic nozzle profile assuming for F^* the area of the cross-section of the nozzle throat so that

$$\frac{F(\delta)}{F^*} = 1 + (2\delta \tan \alpha)^2 \,, \tag{2.83}$$

where 2α is the total opening angle of the nozzle. By substituting into (2.82), an explicit solution can be found for $M(\delta)$, which, for $\delta < 20$, assumes the simplified form

$$M_{\mathrm{hyp}} = 1 + 2.82\delta \tan\alpha \, . \tag{2.84}$$

At this point, all the required elements are available to solve (2.81), which is an initial value problem. If one assumes that at the nozzle throat there is no condensation $N(0) = 1$ (only monomers are present at the beginning of the expansion), (2.81) may be easily solved numerically.

Weiel has shown that this model gives predictions for MgF_2 seeded in Ar that compare well with experimental results with a maximum cluster size of 150 for a nozzle with $\alpha = 5°$ and a throat diameter of 0.29 mm, a crucible temperature of 2200 K and an Ar pressure in the source of 4000 hPa. The model shows also how relevant cluster–cluster collisions could be in such a regime, demonstrating that aggregation is quite relevant in seeded expansion at high dilution rates.

3. Cluster Sources

3.1 Vaporization Methods

The availability of efficient atomization methods is the prerequisite for the realization of an intense cluster source. Another requirement is the control of the parameters of gas aggregation determining the growth rate, and hence the mass distribution of the particles. Once formed, the clusters must be extracted from the source with their original physico-chemical characteristics unperturbed.

In order to produce nanostructured materials with tailored properties, one should also be able to control the parameters of cluster–cluster and cluster–surface interactions which are significantly affected by the kinetic energy of the aggregates.

Different cluster production techniques fulfill several of the requirements cited above; however, an ideal cluster source where all the parameters can be controlled is very difficult to obtain. For example, aerosol sources can produce nanoparticles in bulk quantities, but they have the drawback of a limited control of the particle residence time in the growing zone and on particle collection. On the other hand, supersonic cluster sources not only allow a better control of these parameters, but also favor further cooling during the expansion. Moreover, supersonic and effusive sources are suitable for the extraction of clusters and for the formation of directional beams. Unfortunately, the requirement of pumping systems make the realization of intense supersonic sources a difficult task.

Experimental results suggest that a combination of aerosol techniques with continuous and pulsed beams could overcome many, if not all, of the major limitations and make cluster beams an interesting tool for nanostructured material synthesis.

In this chapter we describe and discuss the experimental techniques most relevant for atomic vapor (plasma) production and how they can be efficiently used to prepare cluster beams. In particular, we discuss the design and construction of sources suitable for cluster beam deposition.

3.1.1 Joule Heating

In many respects the design of a cluster beam source based on a heated oven is identical to that of an atomic beam source. It requires a reservoir where a certain vapor pressure of the material under study must be sustained and a well defined exit opening. In a cluster source it is also necessary to vary the vapor pressure and temperature to induce nucleation: this is usually realized by mixing the vapor with an inert gas at a controlled temperature or by producing a pure or seeded supersonic expansion of the vapor.

In practice the construction of cluster oven sources is complicated by the need for the high temperatures at which they must operate. The choice of suitable materials, their assembly and gas handling place requirements and constraints on the design of a high temperature cluster source that must be satisfied in order to achieve the desired characteristics of the cluster beam in terms of intensity, mass distribution, kinetic energy, etc.

High-temperature crucibles have been widely developed for molecular beam epitaxy (MBE) [3.1]. Nonetheless, high intensity sources for the production of cluster beams have more stringent requirements that deserve some specific considerations. In particular, the vapor pressures used in MBE are typically much lower (of the order of 10^{-3}–10^{-4} mbar). Hence they operate at temperatures significantly lower compared to the sources discussed here that require regimes of vapor pressures two order of magnitude larger. In practice, some of the materials commonly used for evaporating a certain metal in MBE become useless at the temperatures needed in cluster sources, where the same material may become much more reactive and may even become a molten liquid. The formation of alloys and eutectics between the crucible and the evaporating substance may give rise also to very unpleasant and unexpected behaviors that could even seriously damage the source.

Phase diagrams [3.2], even though often inaccurate, can be a guideline in the choice of the crucible material. In Table 1 of Appendix one can find data concerning melting point, boiling point, and temperature needed to operate the source at a useful vapor pressure (1.5×10^{-2} mbar) for materials that can be used or have been used in atomic beam sources. The elements of interest for building crucibles and assembling sources are also reported.

It is important to underline that the crucible should not interfere with the source in terms of reacting with the evaporating material or producing a vapor pressure of its own that would contaminate the beam: sublimation can often be an efficient source of vapors. Under the column "compatibility" of Table 1 (Appendix) we report some of the known problems connected with the materials used for constructing the source, or to be used to produce a beam as a combination of the two. The last column briefly describes some known hazard problems connected with the use of the specified materials. These aspects should not be overlooked since serious problems can arise from the handling of some metals even at room temperature. The list is meant to be a guideline only, so that before designing a new source one should carefully check these aspects. Good sources of information are some chemical catalogs (i.e. Aldrich Catalogue Handbook of Fine Chemicals) and [3.3, 3.4].

Another aspect to be carefully considered in choosing the material for the crucible, is the heating method. The most frequently used are resistive, electron bombardment and induction. A possibility that, in principle, could simplify the source construction is direct heating in which the crucible itself is the active element (resistance or susceptor in the case of induction). The resistivity of the material, and its temperature dependence over the whole operative range of the source, then become critical parameters. Figure 3.1 shows the resistivity [3.5] of some of the best candidates for this approach: graphite, Kanthal, Ni/Cr, Ta, Nb, Mo and W.

The choice of the material for devising the heat element should be based on the temperature regime in which the source will operate and on the lifetime expected. This is affected by the shape and size of the heating element and by the local environment. A detailed discussion on the advantages and disadvantages of the different heating materials is given in [3.1].

Carbon represents one of the best choices since it offers the highest resistivity even at room temperature. Even though its temperature dependence is not monotonous, and it gives rise to a minimum of about $6.4\,\mu\Omega$ m around 1000 K, carbon may still be considered an ideal material even at high temperatures. Another important factor in favor of carbon is that high density graphite is commercially available from different manufacturers in different grades so that one could find the most suitable carbon material for each specific application.

From the point of view of source design there are no particular drawbacks in using carbon since it can be easily machined and offers good mechanical strength so that high quality crucibles can be made which operate reliably up to 2500 K [3.1, 3.6–3.14].

Crucibles made of refractory metals can also be easily constructed by spark erosion or can be easily found on the market. Direct heating at low temperatures is not convenient due to the low resistivity of these materials, as shown in Fig. 3.1.

In a temperature regime typically up to 1300 K, in indirectly heated sources, the use of alloys for the heating elements may be convenient. They offer two main advantages: they can be easily bent and shaped without breaking and show a small variation of resistivity over their working range. Two alloys are most commonly used: Nichrome 80 (80/20 Ni/Cr alloy) and the ones manufactured by Kanthal that offer a resistivity at room temperature of about 108 and $135\,\mu\Omega$ cm respectively. The use of these alloys, at high temperatures, is limited by the non stoichiometric evaporation of Cr that produces a decrease in resistivity. In the case of nichrome this effect determines a maximum operating temperature of about 1100 K while Kanthal, especially if oxidized before use in vacuum, withstands higher temperatures.

Regimes of even higher temperature may be achieved using wires or heating elements made of pure W, Ta or Mo. As already mentioned, even though

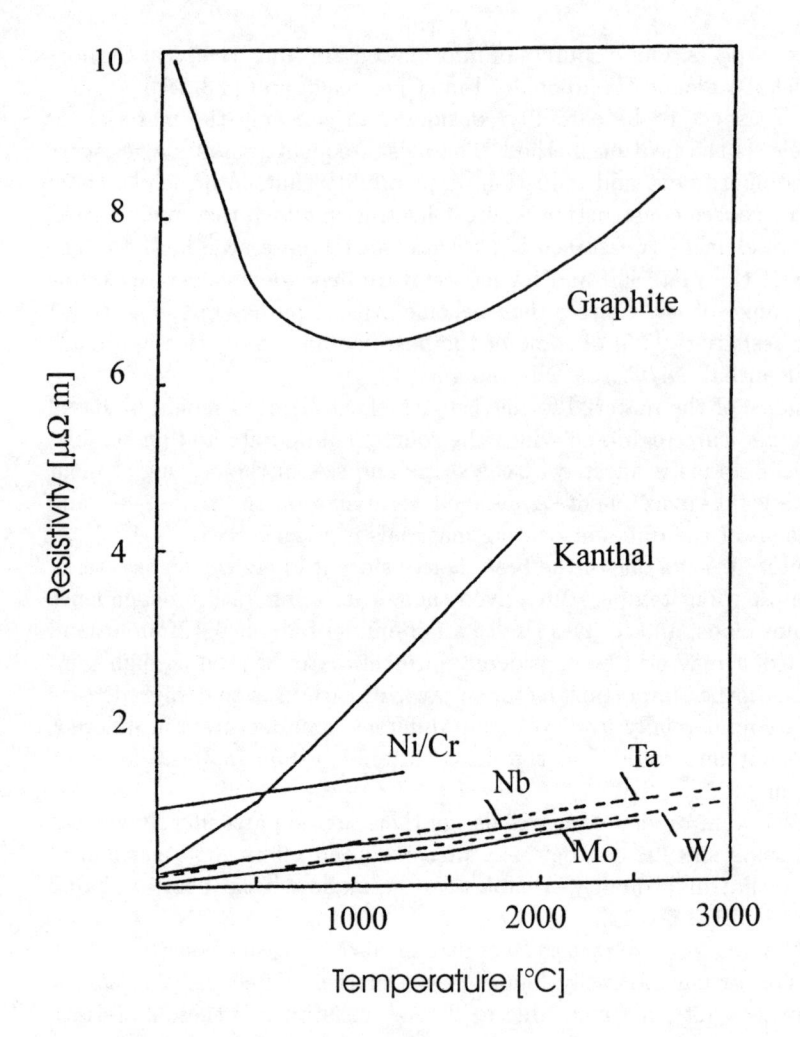

Fig. 3.1. Resistivity of some elements in the range of temperatures useful for crucible construction. (From [3.5])

their evaporation losses are still quite small at very high temperatures (2830 K for W, 2700 K for Ta and 2180 K for Mo) one should be careful about the effects of high temperature reactivity of these metals with the material to be vaporized. Even though it is the most expensive, tantalum is often the heating material of choice since it is not brittle, both before and after heating, it has resistivity and emissivity similar to those of tungsten and the effective maximum operating temperature is only slightly lower than for tungsten. Less commonly used solutions include non-metallic materials such as SiC, which, even though it is brittle, can be machined to the desired design and can be operated up to 1850 K. Unfortunately, several thermal cycles produce a significant increase in its resistivity.

A critical role in the design of metal vapor sources is played by ceramic materials such as refractory oxides (Al_2O_3, MgO, ZrO_2 and complex oxides) that are used as insulators in the form of bushes and washers and by machinable ceramics (such as boron nitride, Macor and Shapal-M) that allow more specifically tailored designs. Some properties of interest here of the most commonly used materials are listed in Table 2 of Appendix.

Alumina and boron nitride are the most widely used ceramics in thermal vapor sources at temperatures up to about 1800 K, since they have high melting points and good resistance to thermal shocks. They are available in a variety of sizes and shapes and can be ground to precise dimensions. At higher temperatures, BN is the material of choice since its excellent thermal properties allow temperatures as high as 3270 K to be reached, where it begins to sublimate. BeO is of interest because its thermal conductivity it is quite high (comparable to metals), even though it is a very good electrical insulator. Unfortunately, its powder is highly toxic so that it requires great handling care. Zirconia, on the other hand, has a limited use at high temperature because it becomes electrically conducting.

At high temperatures one should also take into account the effects due to the vapor pressure, to the reactivity and to the stability of the ceramic materials especially when in contact with the metal to be evaporated [3.15, 3.16]. For example, MgO, as shown in Table 2 of Appendix, generates a vapor pressure of 1.5×10^{-5} mbar at a temperature of 1500 K so that its vaporization is quite sizeable, making its use limited at temperatures lower than 2300 K. The maximum temperature for surface-to-surface stability of BN with carbon is only 2170 K and it drops further to 1770 K with Ta [3.1].

Machinable ceramics are playing an increasingly important role in the construction of ovens since they allow specifically tailored designs. Boron nitride and Macor both have no known toxic effects, are machinable to high tolerances and are compatible with ultra high vacuum environments. Macor is a glass ceramic, containing a mixture of oxides and a few percent of fluorine, that can be used continuously at a temperature of 1070 K (peak value of 1270 K). Shapal-M is an aluminum nitride ceramic offering ultra high purity and it can be used in a nonoxidizing atmosphere at temperatures up to 2170 K. Its thermal conductivity is $100\,\text{W}\,\text{m}^{-1}\,°\text{C}^{-1}$, about two orders of magnitude larger than Macor and five times that of alumina.

The design of a thermal vaporization source based on resistive heating, schematically includes a crucible and a series of radiation shields made of Mo or Ta kept in place by ceramic holders and spacers. In this way, one can reduce radiation losses and related problems such as unwanted heating of parts of the apparatus that in turn could produce vapors that may contaminate the beam itself. At very high source operating temperatures, the outermost shield can be cooled by water or liquid nitrogen. This solution is often needed to reduce the background gas due to the metal vapor that may contaminate the vacuum system. This is an effect that depends strongly on the sticking coefficient between the atom vapor and the surrounding surfaces and can be quite severe for certain species (for example, cadmium) [3.17, 3.18].

In order to avoid partial or total clogging of the source the nozzle should be kept at temperatures higher than the oven (typically 50 K). This can be achieved by separate heating elements that increases the flexibility in controllling the temperatures. An electronic feedback between adequate thermocouples (for example chromel–alumel for temperatures lower than 1400 K and tungsten-tungsten-rhenium or tungsten-rhenium-tungsten-rehnium in the range 1300–2500 K) properly positioned in contact with the oven and with the nozzle, respectively (usually by spot welding them directly or by means of Ta supporting disks) and the current flowing through the heaters will guarantee the needed temperature stabilities. A typical source based on resistive heating is shown in Fig. 3.2.

Electron bombardment can also be used to reach the high temperatures needed to vaporize materials. In these sources, the electrons are accelerated, typically in the range between 2 and 10 kV, and focused directly on the evaporant by electric or magnetic fields. To avoid, or reduce, problems related to the impact of electrons with the vapor (production of excited or ionized species) the heat can be transferred by a radiator that is bombarded by the electrons.

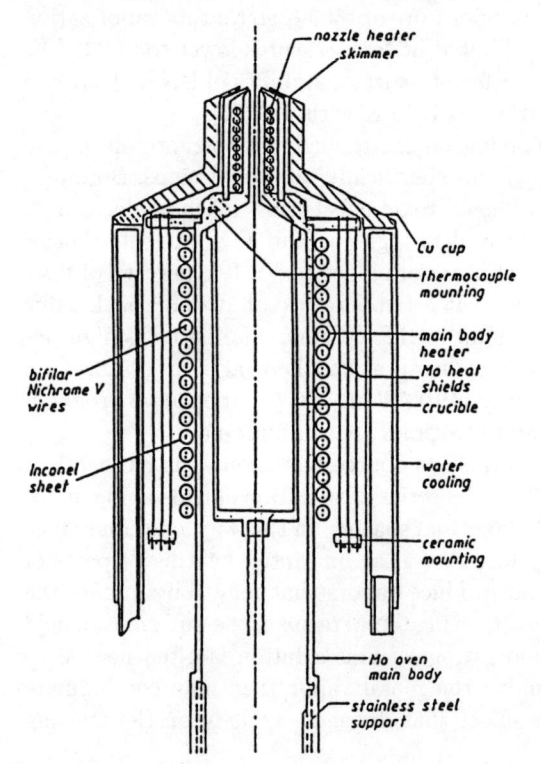

Fig. 3.2. Schematic of a typical metal vaporization source based on resistive heating. (From [3.1])

Extreme source temperatures can be achieved only by inductive heating. This is the case of materials that require temperatures higher than 2700 K to produce a vapor pressure larger then 10^{-2} mbar. As mentioned, a choice that can simplify the construction is a crucible that acts also as a susceptor, otherwise a good thermal contact between the two should be guaranteed. In the choice of materials one should avoid interference between the susceptor and the surrounding components so that the power is dissipated at the right point. The susceptor is surrounded by a work-coil through which the high-frequency (500 kHz to 1 MHz) current circulates. The coil is usually made of copper tubing inside which water circulates. The joule heat produced in the susceptor is proportional to both its resistivity and the induced current. A trade-off should then be achieved between high joule effects and high induced currents: typically, high resistivity materials have lower induced currents. One should keep in mind that the heating of the susceptor depends on the so-called skin depth and on the thermal conductivity. The former is defined by the depth at which the induced current in the susceptor decreases by a factor $1/e$ of the value at the surface in contact with the coil. In order to reduce the heat flow out of the crucible, and since 86% of the power is dissipated into this skin depth, an ideal crucible should have a thickness comparable to the skin depth. In practice, graphite is a good susceptor material even though it has a high resistivity: its skin depth at 1 MHz is 1.8 mm. Tungsten is quite good at high temperatures because it increases its resistivity, while at low temperatures it requires very high power. Figure 3.3 shows a typical set up for such sources.

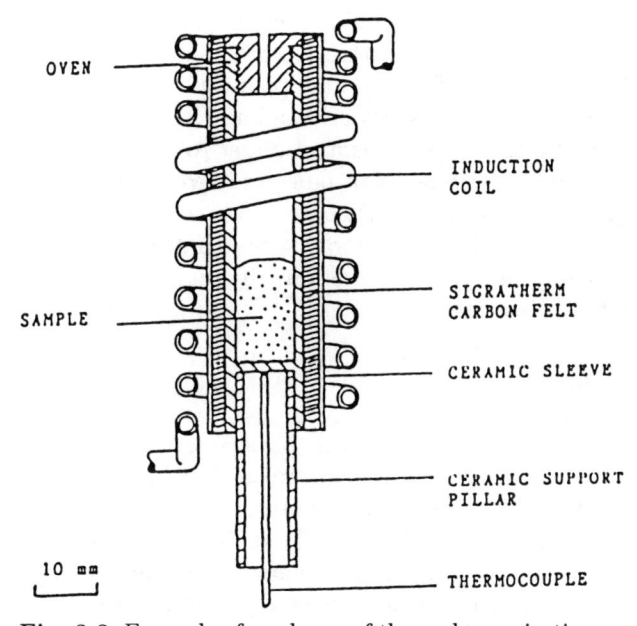

Fig. 3.3. Example of a scheme of thermal vaporization sources based on inductive heating. (From [3.1])

With a correct choice of the materials and an accurate design, induction heating extends the use of thermal vaporization to a large variety of elements, including some refractory materials [3.19] (see Table 1 of Appendix). At these extremes the drawbacks typical of thermal vaporization become quite relevant. The need to heat relatively bulky parts, and the effects of heat flow to the rest of the apparatus induce the mentioned problems of conyamination due to the outgassing, vaporization coming from the hot parts and reactions occurring inside or around the crucible. These effects strongly affect the lifetime and reliability of the source.

3.1.2 Plasma Generation for Cluster Production

When an energetic atom or ion hits a solid surface, other particles such as neutral or ionized atoms, molecules and clusters may be ejected as a consequence of the energy and momentum transferred to the solid. This process is relevant for cluster production and represents an alternative to joule-heated ovens since it can be applied to virtually any material and the sputtered species can be efficiently used as building blocks of aggregates.

Ideally one should be able to tune, at an atomistic level, the particle emission rate (sputtering yield), their density, temperature and velocities, and the energy exchange processes in order to control the cluster formation. In practice this is very difficult due to the large number of parameters involved and the difficulty of characterizing each single step from the ejection of precursors to cluster condensation.

Among the different methods of vaporization based on particle bombardment, plasma discharges deserve particular attention here, since they can be easily and economically coupled to different beam apparatus.

Before plasma discharges, we first discuss in detail another vaporization method based on the transfer of a huge amount of power to a solid surface by a laser pulse. This method, in a sense, can be considered to lie midway between conventional joule-heated oven sources and plasma sources since, via the localized heating induced in the material, one can achieve an efficient vaporization suitable for cluster condensation without the mentioned drawbacks due to high heat dispersion.

3.1.3 Laser Vaporization

Laser ablation of solids produces a plasma with physico-chemical properties suitable for creating structures relevant for microelectronics fabrication, deposition of multilayered structures, and synthesis of superconductor and ferroelectric thin films [3.20–3.22] (Fig. 3.4). By rapid quenching of the plasma, favoring condensation, clusters and nanoparticles can also be produced [3.23–3.26]. Laser vaporization is an interesting alternative to other

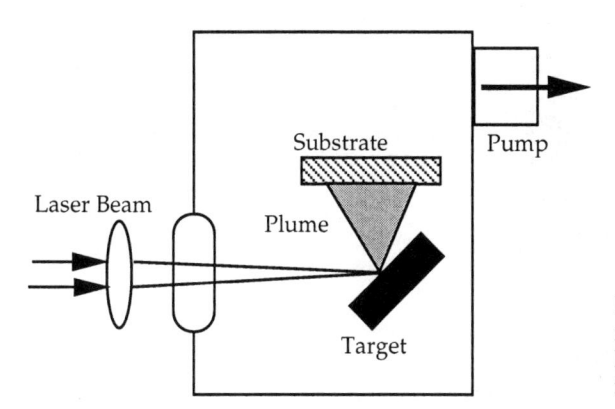

Fig. 3.4. Schematic representation of a laser ablation apparatus for thin film deposition

heating methods: it can generate a high density vapor of virtually any material in a short time interval, the produced vapor is characterized by high speed and directionality.

This technique was originally applied to the production of clusters in molecular beams by Smalley [3.23]. Since this pioneering work laser vaporization has become one of the most common techniques for generating cluster beams of refractory materials. The use of laser vaporization cluster sources (LVCS) for the synthesis of nanocrystalline materials, has only recently been proposed. For this purpose the requirement of intense and stable beams demands an optimization of several parameters which will be discussed in more detail in Sect. 3.3.2. Here, we describe and discuss the basic features of laser ablation of solids that are preliminary to the construction and operation of LVCS.

The production of clusters and fine particulate is due to the condensation of the plasma produced by laser irradiation. Species emitted from the irradiated surface can interact mutually, with the target and with the terminal part of the laser pulse [3.27] (Fig. 3.5). The presence of a background gas can favor the thermalization of the hot vapor expanding at high velocity and the formation of metastable complexes. The background gas may also de-excite the complexes, forming the seeds for the formation of larger species.

The kinetics of aggregation of the plasma produced by laser ablation and the characteristics of the resulting aggregates (density, size distribution, structure, etc.) are influenced by the quantity of ablated material, the type of ejected particles, the nature and the dynamics of the plasma plume, and the interaction with the background gas. Laser power density, pulse duration, wavelength, as well as the thermodynamic and optical properties of the target affect the laser–solid interaction and the nature of the ejected products.

Laser light absorption by a solid can be modeled by the relationship:

$$I = (1 - R) I_0 e^{-\alpha x} \qquad (3.1)$$

where I_0 is the laser fluence (W cm^{-2}), I the fluence at a distance x beneath the surface located at $x = 0$, α is the absorption coefficient and R the re-

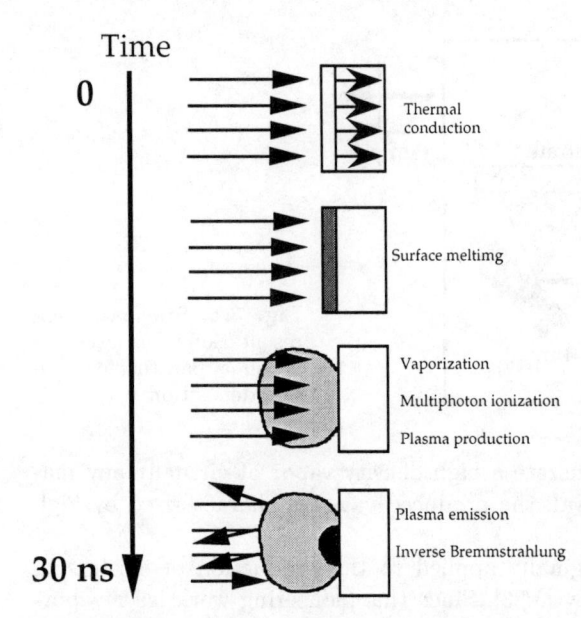

Fig. 3.5. Schematic representation of the laser–target interaction. (From [3.27])

flectivity [3.28]. The target material is also characterized by a specific heat C_V, a density ρ, and the thermal conductivity K. The thermal diffusivity $D = K/C_V\rho$ determines the thermal diffusion length after a time t of laser irradiation: $L_{th} = (2Dt)^{1/2}$ [3.29]. For a laser power density Q_0, the heated layer has gained an average thermal energy $Q_0 t/(Dt)^{1/2}$. When this energy exceeds the sublimation energy of the solid in a time shorter than the duration τ of the laser pulse, evaporation takes place [3.30]. The onset of different regimes depends on the energy absorbed by the material, the amount of laser power absorbed, and the other parameters being constant. In the case of metals, the absorption is related to the emissivity $\varepsilon = 4n/[(n+1)^2 + k^2]$ where n and k are the real and the imaginary part of the refractive index. In Table 3.1 we report the value of emissivity for various metals. It can be seen that absorption increase in the visible region of the spectrum [3.30].

Using lasers with pulse widths in the nanosecond range, the minimum power density for the vaporization of metals is 10^9 W cm^{-2}. For semiconductors and insulators the required power density is lower due to the lower reflectivity. The absorbed power is a strong function of wavelength [3.29].

The formation of a vapor requires the presence of a liquid phase and the establishment of an equilibrium between solid, liquid and gas phases. The vaporization process creates a recoil pressure that pushes the vapor away from the target and expels the liquid (Fig. 3.6). The material is thus removed both in the gas and in the liquid phases (droplets).

The rate of material removal depends not only on heat transfer but also on the gas dynamics of the vapor [3.28, 3.31]. The phase (vapor or liquid) of the ablated material is strongly dependent on target material, on laser fluence

Table 3.1. Values of emissivity at 20 °C for various metals at different laser wavelengths (adapted from [3.30])

Metal	Emissivity			
	Ar$^+$ (500 nm)	Ruby (700 nm)	Nd:YAG (1000 nm)	CO_2 (10 µm)
Al	0.09	0.11	0.08	0.019
Cu	0.56	0.17	0.10	0.015
Au	0.58	0.07	–	0.017
Fe	0.68	0.64	–	0.035
Pb	0.38	0.35	0.16	0.045
Mo	0.48	0.48	0.40	0.027
Ni	0.40	0.32	0.26	0.03
Nb	0.58	0.50	0.32	0.036
Pt	0.21	0.15	0.11	0.036
Ag	0.05	0.04	0.04	0.014
Ta	0.65	0.50	0.18	0.044
Sn	0.20	0.18	0.19	0.034
Ti	0.48	0.45	0.42	0.08
W	0.55	0.50	0.41	0.026
Zn	–	–	0.16	0.27

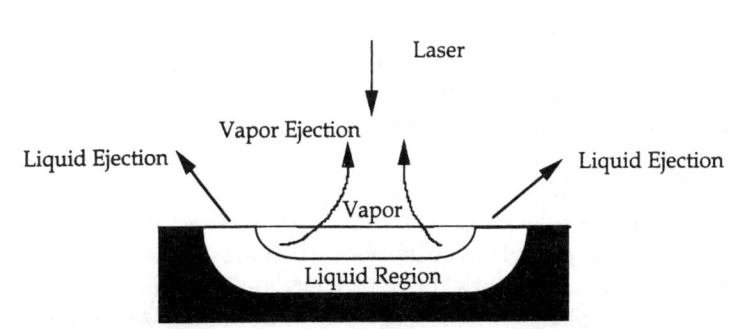

Fig. 3.6. Schematic diagram of the process of material removal. (From [3.31])

and on surface roughness, evolving during irradiation [3.32]. These aspects are very important since the expansion of the plasma plume, its interaction with a buffer gas and the ejection of droplets of molten material, strongly affect the degree of condensation, so that they should be taken into account in designing a LVCS. In particular we underline that the characteristics and the quantity of the ejected material vary with time due to the evolution of target morphology under the irradiation spot.

At high fluences a substantial vapor pressure develops above the irradiated region causing the onset of nucleation and backflow. A large change in density associated with vapor nucleation gives rise to a violent expansion of the superheated liquid interface: the evaporated atoms can then acquire a considerable thermal energy. For a solid formed by different elements, the evaporation takes place without changing the original chemical composition of the target, regardless of the differences in the chemical enthalpies of the various elements [3.33].

The laser-induced vapor plume evolves very rapidly and it can further absorb part of the laser radiation. In these partially ionized gases, light can be absorbed by thermally excited atoms (bound–free absorption) or by ions (bremsstrahlung absorption) [3.22]. With increasing irradiance, the temperature and the enthalpy of the gas increase, giving rise to an even larger absorption. This positive feedback favors the formation of a plasma from the evaporating target. Once the gas is fully ionized, the absorption is dominated by bremsstrahlung processes.

Material ejection is then a process affected by laser–solid, laser–plasma and plasma–surface interactions. The ejected particles contain a mixture of ions, electrons, hyperthermal neutral atoms, molecules and larger species in electronic excited states, UV photons, and even soft x rays [3.27].

The plume of evaporated material is cone-shaped and has a highly forward peak distribution described by a cosine law $\cos^n(\theta)$ with $8 \leq n \leq 12$, where θ, measured from the target normal, is the light angle of incidence. In addition to this, a second $\cos(\theta)$ distribution is observed indicating the presence of different mechanisms for material removal [3.27].

The velocity distribution of the particles ejected from the surface is very complex and, in general, assuming a thermal distribution, it cannot be described by Maxwell–Boltzmann statistics [3.27]. The number of species arriving at a detector at time t can be described by:

$$N(t) = Kt^{-4}\exp[-v^2/(2kT/m)]\,, \tag{3.2}$$

where K is a scaler, v the velocity of a particle with mass m and temperature T. The typical time-of-flight distributions observed can be fitted by multiple distributions generated with different characteristic temperatures. This behavior has been explained by the formation of a Knudsen layer above the target surface. The variety of collisional processes due to the interaction of the particles in the plasma can give rise to a non-Maxwellian velocity distribution [3.28] (see below).

Laser ablation in an atmosphere is of particular relevance for efficient cluster formation. Compared to vacuum, ablation in a buffer gas differs for three major features [3.21]:

   (i) production of a shock wave;
   (ii) reduced extent of the plasma plume due to confinement;
   (iii) altered expansion dynamics of the plume.

Shock waves are formed by the pressure exerted on the background gas by the rapidly expanding material ejected from the target. As the shock expands, the expansion velocity decreases with increasing distance from the target [3.21]. This type of non-steady shock wave is referred as a blast wave and it is produced when a large amount of energy is deposed in a small volume with an atmosphere to support the wave. Blast-wave velocities as high as Mach 27 have been observed in iron ablation experiments with an atmosphere of 10 torr of argon [3.21].

A buffer gas alters the mechanism of plasma production and plasma ignition. In particular, for the expansion of a plasma plume, its presence implies deceleration of the ablated species, formation of turbulence, condensation, redeposition of the ablated particles, and reaction of the ablated species with the background gas [3.21]. Moreover, a considerable amount of laser energy can be coupled to the buffer gas rather than to the target, reducing the quantity of the ablated material compared to ablation in vacuum.

The velocity of ablated species in vacuum is of the order of $1 - 2 \times 10^6$ cm s^{-1}: collisions with the buffer gas decelerate the species from the initial velocity down to 500–3000 cm s^{-1}. in approximately 1 ms after the ablation event [3.21]. In the case of laser ablation of copper, the average directed velocity of the ablated particles after 100 μs depends strongly on the gas pressure. Velocity values span a range from 2×10^4 cm s^{-1} with 10 torr of He to 8.6×10^3 cm s^{-1} with 100 torr [3.21]. Most of the condensation is observed to take place after 100–200 μs so that the vapor is almost completely decelerated before condensation starts.

Turbulence and deceleration both favor mixing and formation of high density regions where condensation starts. As already described in Chap. 2, the formation of clusters is related to the formation of small condensation nuclei. Excited dimeric species are produced by collisions between ablated atoms that occur with at a rate proportional to the square of the target atom density. The buffer gas favors the formation of high density regions and also thermalizes the energy of the excited dimers. Vortex formation further improves the condensation efficiency. Depending on the degree of excitation of the dimers, a substantial cooling can occur before the addition of atomic species and the growth of clusters [3.34].

3.1.4 Glow and Arc Discharges

Plasma discharges are of particular relevance for cluster formation. Schematically a discharge consists of a voltage supply that drives current through a low pressure gas between two conducting plates or electrodes. The gas breaks down to form a weakly ionized plasma (Fig. 3.7).

Charged particles in the plasma acquire high kinetic energies and collide with the neutrals in the gas, causing the formation of very reactive species. Moreover, material ejection from the plates or electrodes can also take place. In this way the electrode material acts as a feedstock of particles injected in the plasma [3.35].

The discharge of a capacitor into a circuit incorporating a gap between the electrodes is usually known as "gas discharge" [3.36]. Among the various gas discharges one of the most studied is the glow discharge which is a self-sustaining discharge characterized by the emission of electrons due to positive ion bombardment on a cold cathode [3.36]. A glow discharge can be ignited if

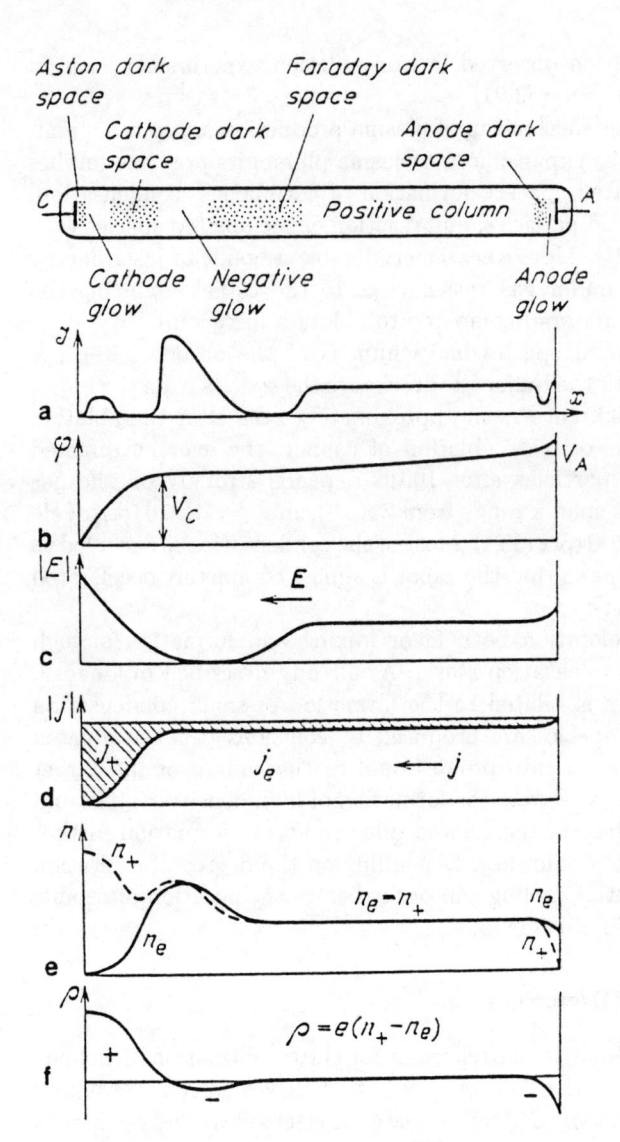

Fig. 3.7. Glow discharge in a tube and distributions of: (**a**) glow intensity, (**b**) potential φ (**c**) longitudinal field E, (**d**) electronic and ionic densities j_e j_i, (**e**) charge densities and (**f**) space charge. (From [3.36])

a layer of large positive space charge is formed at the cathode. An electrically neutral plasma region is present between the cathode layer and the anode if the electrode spacing is sufficiently large.

One of the most widely used plasma discharge configurations consists of a vacuum chamber containing two planar electrodes separated by a few centimeters and driven by an rf power supply (Fig. 3.8). Substrates for thin film deposition are placed on one electrode. Feedstock material is introduced in gaseous form or sputtered from a target [3.35].

Direct-current glow discharges are used for thin film deposition, as metal sputtering sources. As rf diodes they present high sheat voltages and low

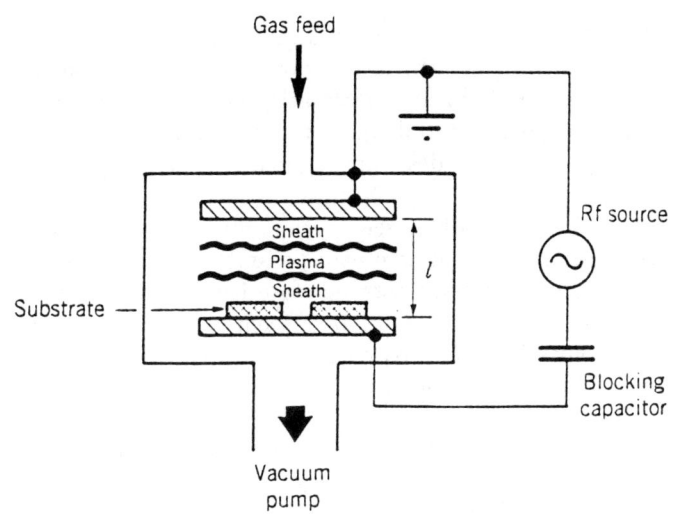

Fig. 3.8. Schematic representation of a symmetric rf diode in a parallel plane geometry. (From [3.35])

plasma densities. Ion bombardment leads to cathode sputtering, which can be used for deposition of thin film or cluster condensation. In the so-called planar magnetron configuration, higher current densities can be obtained by applying a dc magnetic field at the cathode to confine secondary electrons produced by ion bombardment.

An interesting pulsed source of highly ionized metal plasma is the arc, consisting of a discharge between two conductive electrodes (Fig. 3.9). The arc regime is characterized by a relatively low cathode potential fall (of the order of the ionization potential of the atoms, ~10 eV) in contrast to glow discharge where the potential drop is several hundred volts [3.36]. This is due to different mechanisms underlying cathode emissions: arc cathodes emit a large electron current as a result of thermionic, field electron and field emis-

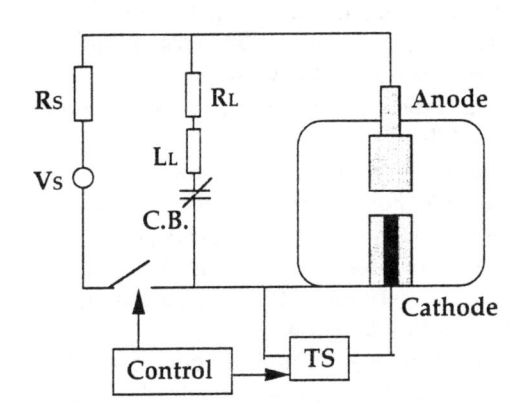

Fig. 3.9. Arc discharge apparatus. (From [3.37])

sion [3.36]. Arc discharges are characterized by large currents $I \sim 1\text{--}10^5$ A, causing a considerable heating of the cathode either of the whole or in a small area for short time intervals, and inducing vaporization of the electrode material and erosion. The vaporized material, ejected under various forms and forming a plasma between the electrodes, can undergo recondensation into clusters and small aggregates.

The phenomenology of the arc discharge is very large and presents a wide variety of different aspects depending on the parameters that influence the emission process: current intensity, electrode materials and geometrical arrangement, plasma state, medium in which the discharge takes place, etc. The classification of arc types is usually based on the characteristics of the electrode processes and, in particular, on cathode operation [3.36, 3.38]. For our purpose we can divide the arcs into two categories: vacuum arcs and pressure arcs. Vacuum arcs are by far the most studied; they can be divided into:

(i) cathode spot arcs or cold cathode arcs;
(ii) hot thermionic cathode arcs.

The cathode-spot arc is a high current, low voltage electrical discharge in which a substantial portion of the conducting medium consists of ionized cathode material generated in localized high current regions on the cathode surface known as cathode spots. The current flows through one or more spots that appear and disappear with a rapid and random motion on the cathode surface. The cathode material is evaporated from these regions. Cathode-spot arcs are used for deposition of thin films and hard coatings [3.39–3.41].

Hot thermionic cathodic arcs are characterized by cathodes heated as a whole to very high temperatures (~ 3000 K) so that the high current results from thermionic emission. The arc is anchored to a fixed large area spot on the cathode, current densities are of the order of $10^2 - 10^4$ A cm^{-2}. These arcs can be obtained with refractory materials (carbon, tungsten, molybdenum, etc.) which can withstand high temperatures.

The anode usually acts as a collector of the particles emitted from the cathode, however, under certain conditions, anode spots can be formed. In this case the anode surface is an additional and often dominant plasma source: anode spots are observed if the current exceeds 10 kA, in the case of hot thermionic cathode arcs, and when the vapor pressure in the vicinity of the cathode exceeds 1–10 torr [3.38, 3.41, 3.43].

Each cathode spot produces a high velocity jet of highly ionized material with kinetic energies in the range 50–150 eV. For refractory materials and high ionization states these energies are usually larger. Multiple ionization is quite common except for graphite and high vapor pressure materials. The plasma jet peaks in the direction normal to the cathode surface and its angular distribution can be approximated by a cosine function [3.41].

Depending on the cathode material, average cathode temperature and ratio between electrode diameter and separation, a considerable ejection of

Table 3.2. Cathode spot temperatures and estimated vapor pressures of vacuum arcs of various metals. The column marked T_c is the measured cathode spot temperature. The column marked T_c/T_b is the ratio of T_c to the boiling point of the material. The column P_c is the vapor pressure in front of the cathode spot estimated on the basis of T_c. (From [3.42])

Metals	T_c (K)	T_c/T_b	P_c (atm)
Au	3620	1.17	5.9
Ag	2770	1.15	4.5
Cu	2570	0.905	0.4
W	5210	0.900	0.2
Mo	4860	0.985	0.6
Ta	9550	1.73	95
Ti	5570	1.56	60
Pd	3630	1.10	3

clusters and macroparticles can also take place. Macroparticles are considered as a significant source of vapor production both during the arc and in the post-arc recovering period [3.44, 3.45]. Velocities of cathode-produced molybdenum macroparticles have been determined in the range of 400–700 m s^{-1} in vacuum arcs with 1 kA current peak and 30 ms duration [3.44]. The macroparticle emission rate is an increasing function of both the arc current and the average cathode surface temperature. A systematic characterization of erosion and ionization in the cathode spot region of vacuum arcs for different materials can be found in [3.46]. We report, in Table 3.3, a characterization of erosion of different cathode materials.

Table 3.3. Experimental conditions for erosion data with cathode of various metals. (From [3.46])

Cathode	Arc Voltage (V)	Electrode spacing (cm)	Arc current (A)	Erosion rate, X ($\times 10^{-4}$ g C^{-1})
Cd	11	1.7	100	6.55
Zn	13	1.7	105	2.15
Ag	20	0.1	105	1.5
Cr	18	0.2	100	0.4
Fe	20	0.2	200	0.73
Ti	20	0.2	100	0.52
C	20	1.7	100	0.17
C	20	0.2	100	0.16
Mo	28	1.7	200	0.47
W	25	0.2	250	0.62

As in the case of laser ablation, the presence of a buffer gas surrounding the electrodes influences the plasma expansion from the electrode spots and also changes the behavior of the arc and the role of the electrodes (appearance of anode spots). In low-current arcs the plasma is confined, by background gas

[3.47,3.48], into a smaller volume compared to the vacuum case. Observations carried out on copper plasma produced in He, Ar, SF_6 at pressures up to atmospheric show that a shock wave is formed during the plasma expansion and it acts as a boundary region. The erosion rate remains approximately constant in vacuum up to 10^{-2}–10^{-1} torr, then falls rapidly by a factor of 15 in the range 10^{-2}–10^{1} torr, depending on the type of buffer gas [3.49, 3.50]. Cathode spot erosion in the presence of a background gas is reduced compared to vacuum, on the other hand, anode spot erosion may increase with increasing pressure. In the case of carbon electrodes, the individual cathode spots group into thermionically emitting single hot spots [3.47].

3.2 Continuous Sources

3.2.1 Effusive Joule-Heated Gas Aggregation Sources

This production method is based on the idea of allowing the cluster to grow from a monomeric vapor. In general, the most critical parameter affecting the source design and the degree of experimental difficulties is the average bond energy ε per monomer in the cluster. In the case of weakly van der Waals bonded clusters ($\varepsilon < 0.3$ eV atom^{-1}) the monomers are typically in the gas phase at room temperature, or can be easily vaporized, on the other hand they require strong cooling rates for an efficient production of clusters. More strongly bonded species, with larger average bonding energies, give rise to quite different cases. The following species would require specific source designs: hydrogen bonded substances with $0.3 < \varepsilon < 0.5$ eV atom^{-1} such as $(H_2O)_n$, $(HF)_n$; intermediate bonded species, with $0.5 < \varepsilon < 2$ eV atom^{-1} such as Na_n, K_n, K_nMg_m, $(NaCl)_n$; refractory materials, with $2 < \varepsilon < 4$ eV atom^{-1}, such as C_n, W_n, Si_nC_m, Ti_n.

The thermal vaporization process becomes more and more difficult with increasing ε, so that different methods including laser vaporization and different kinds of plasma discharges have been envisaged. On the other hand, the cooling rates needed to efficiently produce clusters of elements with larger ε become less stringent so that there is no need to seed in large fluxes of cooling gases and the vacuum equipments can be less elaborate.

As discussed in Chap. 2, high yield, effusive formation of clusters requires non equilibrium conditions. In thermodynamic equilibrium only a minimum amount of atoms can be organized in clusters while the most probable states are gaseous or condensed phases [3.51]. This is why the existing methods for generating clusters are all based on the expansion and non equilibrium cooling of a vapor generated in different ways. The vapor expansion produces a drop in temperature and, below a certain temperature, the vapor pressure will exceed the saturated vapor pressure. At this point the nucleation process starts and is accompanied by the formation and growth of clusters. The process will end when the density drop due to the expansion reaches the point where the collisions needed to produce and grow the clusters are no longer present. The formation of aggregates of a certain size is therefore effective only if the expansion process lasts for a time longer than the typical time needed for the formation of the required cluster.

Another aspect that one should keep in mind is that the formation and growth of the clusters release energy that will heat the beam so that, in principle, it is impossible to transform an expanding gas or vapor in a beam of only clusters, unless the heat of formation is extracted efficiently. The use of buffer gases, usually inert, is a good solution to this problem.

The construction and engineering of cluster sources have to comply with all these requirements by combining an efficient production of the vapor with a slow quenching in an inert gas.

Knudsen sources are, in principle, the simplest type of beam source and they have been developed to produce atomic beams of thermally vaporizable substances. Since the vapor is kept in equilibrium in an oven, it is dominated by atoms but it also contains a small quantity of small clusters and it is usually allowed to effuse in a low pressure region ($\leq 10^{-6}$ mbar). In this way the original composition in the oven is maintained and the loss of material does not affect the vapor equilibrium inside the oven. This is achieved by working in the regime in which no collisions occur while the particles leave the source, i.e. by keeping the oven aperture smaller than the mean free path in the gas.

Typically, Knudsen cells operate at pressures around 1 torr and with orifice diameters of roughly 1 mm, and with these conditions very small cluster intensities are expected, as discussed in [3.52, 3.53]. Although these kind of sources have no practical use in synthesizing thin films from cluster precursors, they have a relevant role in the study of fundamental properties of small clusters such as bond dissociation energies that have been measured for different metals [3.54], ionic and intermetallic compounds [3.55]. In Knudsen cells, the aggregates are formed in an equilibrium state and therefore their composition is well known. Unfortunately, this is often only an assumption since the equilibrium vapor composition is perturbed by effusion in vacuum even at Knudsen numbers $0.1 < K_\mathrm{n} < 1$.

On the other hand, the methods to produce the vapor in Knudsen cells are very similar to those used in gas-aggregation sources with the appropriate modification that will be described in the following paragraph. Hence, some of the technical considerations and problems related to the materials used apply to both of them.

In order to produce a stable and reliable beam the following basic requirements should be kept in mind in the design of the source:

(i) intensity and stability;
(ii) effective operating period (time between two following maintenance or recharges);
(iii) cleanliness and contamination.

54   3. Cluster Sources

Intensity and stability in time and space are strongly affected by the design, the construction and by the materials used. Partial or total clogging of the source opening, thermal instabilities, alloying, melting and boiling are typical processes that can dramatically affect the performance of the source and therefore particular care should be devoted in the design to avoid or minimize them.

The effective operating period depends essentially on the volume of the crucible but constraints due to space, safety and materials requirements strongly limits the duration of the source. Together with recharging, the maximum continuous operating time is limited by several maintenance operations such us unclogging the source and cleaning the system, including collimators and vacuum chambers. An efficient and reliable source should make recharging and cleaning as easy and fast as possible by a modular design that facilitates mounting and dismounting and does not require time-consuming alignment procedures.

Designing Gas-Aggregation Sources. The presence of clusters in a standard effusive beam, produced by thermal evaporation, is residual and therefore the gas-aggregation process should be activated by non equilibrium cooling in a condensation cell. Basically, the idea is to mix the metal vapors with a cold buffer gas (usually a noble gas) and then let it expand in a high vacuum system through a nozzle. The parameters that play a major role in the condensation cell design are: inert gas pressure and temperature, inert gas type, oven temperature, and oven–nozzle distance. The condensation process can be strongly affected by these parameters so that only a careful control of them can guarantee both stability of the cluster size distribution in the beam and the ability to tune it over a wide range of cluster sizes. A schematic view of a typical condensation chamber is shown in Fig. 3.10.

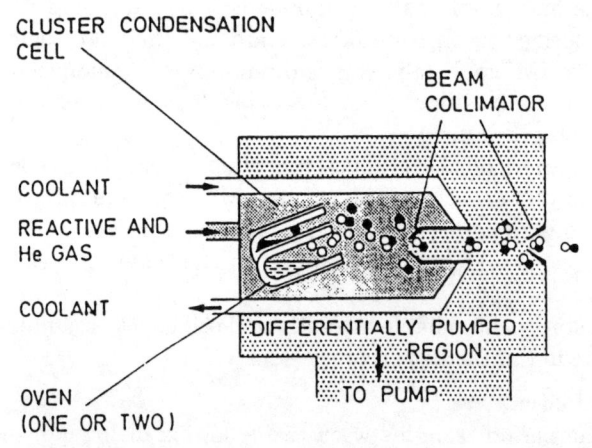

Fig. 3.10. Schematic view of a typical condensation cell of a gas-aggregation source. (From [3.56])

As far as the oven is concerned the considerations discussed in Sect. 3.1.1 apply, although one should consider quite carefully that the oven is inside the condensation chamber and therefore all the mentioned problems related to the high temperature of the crucible (outgassing, heat losses, reactivity, etc.) can affect cluster formation even more dramatically. The production of clusters of mixed metals complicates the situation, since, in this case, one should use more than one oven in the same condensation cell, and maintain the ability to control carefully and independently the temperature of each of them. The relative distance between the ovens, their relative orientation and the distances from the nozzle are hence other parameters to be adjusted to adequately control the stoichiometry, intensity and size distribution of the cluster beam.

Highly efficient generation of metal clusters was first achieved using a gas-aggregation source by Sattler et al. [3.57] (Fig. 3.11). A Knudsen oven is inserted inside a condensation cell cooled to liquid nitrogen temperature. The cooling is obtained by a He partial pressure that fills, via a gas inlet and a valved vacuum system, the condensation cell, thus allowing the nucleation of the clusters. As usual the beam is formed by the effusion of the vapors out of the cell through a collimator of 1.5 mm diameter, a differential pumping stage and a second collimator.

The performance of the source has been tested on Sb, Bi and Pb showing significant cluster intensities with sizes up to several hundreds. Fuchs et al used this kind of source to produce Sb cluster beams with narrow cluster size distributions (down to 5% full width at half maximum) centered at a

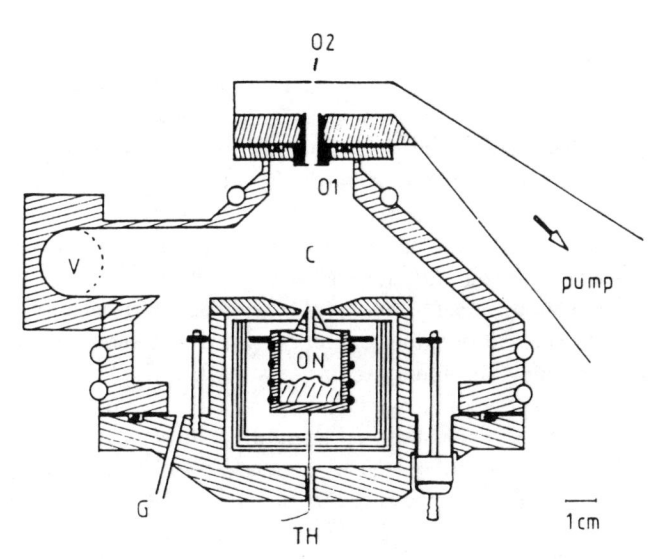

Fig. 3.11. Schematics of the gas aggregation source of the Sattler type: ON, oven; C, condensation cell; O1, O2, collimators; V, valve for vacuum connection; G gas inlet; TH, thermocouple. (From [3.57])

few hundred atoms per cluster [3.58] and at one or two thousand atoms [3.59] without the presence of the molecular species. These different regimes have been achieved by using He as quenching gas to operate up to a few hundred atoms per cluster and Ar to produce beams of larger ones. This aggregation type of source is capable of producing deposition rates ranging from 5×10^{-3} nm s^{-1} up to (for small clusters) 1.7×10^{-1} nm s^{-1} that is with fluxes up to 10^{13} cluster cm^{-2} s^{-1} [3.58].

Slightly different designs are used by Zimmerman et al. [3.60] and Goldby et al. [3.61] allowing the preparation of beams of clusters including up to 10^5 atoms or even larger. The basic design is similar to the one shown in Fig. 3.11. In both cases the major difference from Sattler's design is that the inert gas flows over the oven and is in line with the skimmer. In Martin's source [3.60] the oven is mounted on a linear feedthrough allowing a fine control of the distance from the nozzle. The regime of operation is classical: the oven is regulated to a temperature such that the vapor pressure of the evaporant is about 0.1 mbar while the inert gas pressure in the condensation cell is typically maintained in the range 1 to 3 mbar by a calibrated needle valve. Therefore the cell works in a flow regime (typically 100 mbar l min^{-1} at 90 K) with an average flow velocity of about 10 m min^{-1}. As for Sattler's source the nozzle is conical with a sharp rimmed opening on the top of 3 mm diameter facing the condensation cell while it ends in a tube 30 mm long and 4.5 mm diameter on the opposite side. This construction reduces the build up of condensed material on the nozzle so that a longer useful time between two maintenance cycles of the source is achieved. Since the gas flow increases only slightly shortening the tube length, most of the pressure gradient is located at the nozzle entrance. If a short nozzle is used the total cluster signal decreases sharply due to a less efficient collimation.

As stated before, the temperature of the oven is an important parameter that controls the clustering process since it strongly affects the vapor pressure of the evaporant and the extension of the vapor cloud into the cell. Figure 3.12 shows, as an example, the mass distribution of Na clusters produced at five different oven temperatures in a constant He pressure (1 torr).

The pressure of the inert cooling gas is an important parameter since it changes the cooling rate into the cell and the flow regime over the crucible. When increasing the pressure an effect that should be taken into account is the rise in velocity, and therefore in kinetic energy, that the clusters will experience. This acceleration may become quite important when the pressure is large enough to produce a supersonic expansion that, even if moderate, would strongly affect the source performance. Figure 3.13 shows the mean velocity of Na clusters as a function of the size produced by Martin's gas-aggregation source at different He gas pressures in the condensation cell. A general trend observed, for given source conditions, is a decrease of the mean velocity as the cluster size increases. This effect, usually called "velocity slip" (see Sect. 2.1.2), is caused by the acceleration that the large slow clusters experience due to the collisions with the faster flowing inert gas atoms. The

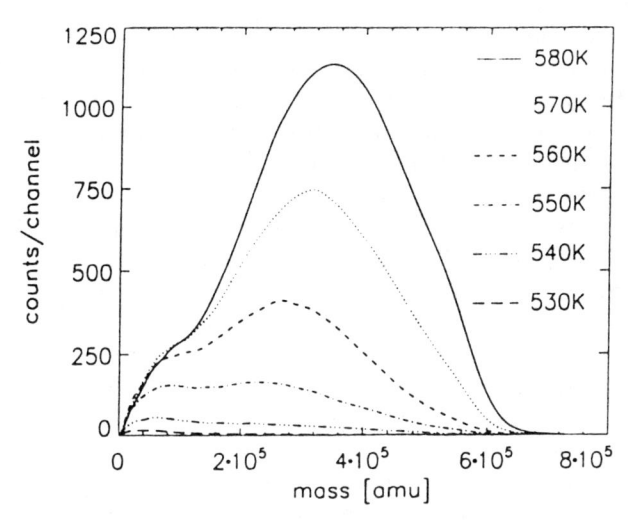

Fig. 3.12. Mass distribution spectra of a beam of Na–clusters produced in a condensation cell at different oven temperatures in a He buffer pressure of 1 torr. (From [3.60])

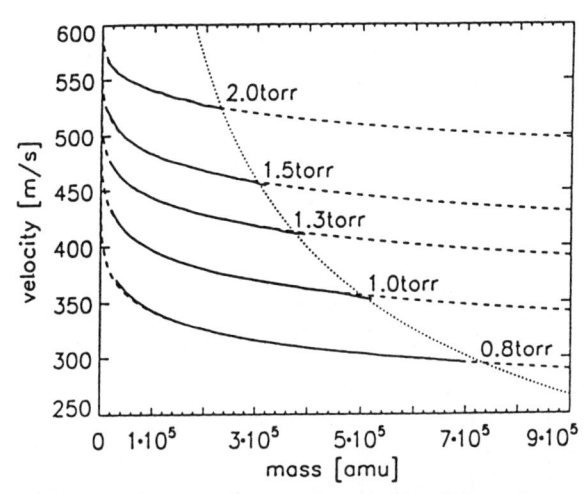

Fig. 3.13. Mean beam velocity as a function of cluster size for Na clusters produced by a gas-aggregation source at different He pressures in the condensation cell. The dotted line represents the ion optics cut off due to the limits at 330 eV of the ion optics in guiding the clusters into the mass spectrometer. (From [3.60])

acceleration region can be assumed to be the length of the nozzle. The effect can easily be understood in the framework of rigid hard sphere collisions. In this model the cross-section of a cluster increases with the square of the radius while its weight grows with the cube so that the smaller cluster are accelerated more readily.

In the limit of very small clusters, the final velocity will be very close to that of the He atoms reaching the regime of zero velocity slip. Due to the different velocity of clusters with different sizes and to its dependence on the source parameters, great care should be used in characterizing cluster beams by ensuring that the detection system maintains the same efficiency over the entire range of kinetic energy spanned (Fig. 3.13).

The nature of the inert gas used in the condensation cell strongly modifies the aggregation process because of the differences in cooling efficiency. Figure 3.14 shows mass spectra of Cs cluster beams formed in a gas-aggregation source using different He/Ar mixtures in the condensation cell. The relative concentration varies from pure He to pure Ar while the oven temperature (420 K) and the total pressure (1 torr) are fixed. It is apparent that the presence of Ar would shift the mass distribution towards larger clusters since its larger mass would increase the cooling rate in the clustering region. One should also keep in mind that, at a given gradient pressure, the Ar flow would be smaller than the He so that the presence of Ar would reduce the velocity and kinetic energy of the clusters in the beam.

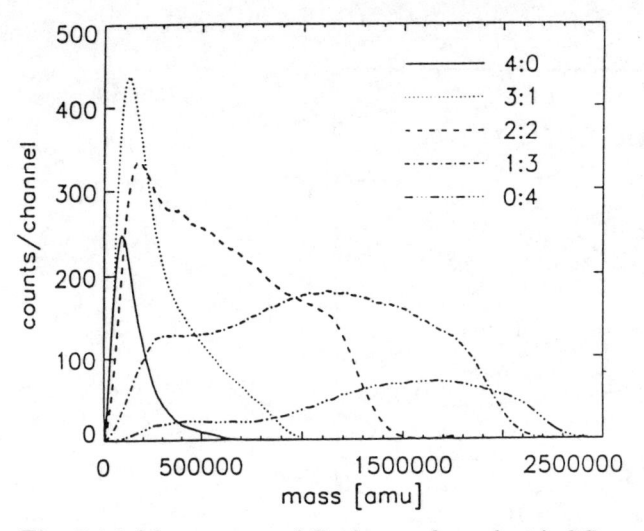

Fig. 3.14. Mass spectra of Cs clusters formed with different He/Ar mixtures. The spectra are labeled with the ratios of He to Ar partial pressures (total pressure 1 torr, oven temperature 420 K). (From [3.60])

As mentioned in Sect. 3.1.1, gas-aggregation sources operating at temperatures high enough to produce clusters of transition metals should be designed using particular care in constructing the crucible. Baker et al. [3.62] have achieved temperatures up to 2020 K with crucibles made of alumina and as high as 2170 K with zirconia. The heater is machined out from graphite while a W/Rh thermocouple measures the temperature. To solve the problem of the excessive heat needed to keep the crucible at this high temperatures the crucible is carefully surrounded by two molybdenum shields. To ensure an efficient cooling to produce enough clustering, the source is build in two stages. In the first one, very similar to Martin's aggregation source, the condensation cell, cooled by water, is operated in an inert gas pressure of about 10 mbar obtained by a capillary bringing the gas flux over the crucible. The mixtures of the metal vapor and of the inert gas passes then through a collimator of a 1.0 mm diameter in a second condensation cell at a lower pressure, typically 0.45 mbar, were it is further cooled to liquid nitrogen temperature. A standard skimmer would finally define the beam that would enter a region of high or ultra high vacuum. The major goal of this design is to produce a continuous change of the temperature of the inert gas between the region of the crucible up to the skimmer so that this whole region would be active for nucleation. The source, specifically made for producing Fe and Mn cluster beams, has not been characterized by mass spectrometry but rather by taking TEM micrographs of the deposited species. The authors show that cluster distributions centered around a few nm are easily achieved at intensities high enough to grow thin films (vapor pressure of the metal in the source typically between 0.1 and 1 mbar). An ionizer and a quadrupole mass filter are used to produce ion cluster beams with narrow size distributions.

In the source designed by Goldby et al. [3.61] the basic idea is to combine the effects of gas-aggregation with those of a supersonic expansion. A cooled stainless tubing (1 mm diameter) brings the inert gas close to the crucible and directed toward the nozzle. The crucible is tilted towards the nozzle in order to maximize the throughput and its temperature can be raised to 1670 K. The operating source conditions are such that the expected terminal Mach number for the expanding He beam would be 2.6, so that the final average speed of the atoms should be $1460 \, m \, s^{-1}$. Since the resulting beam is only mildly supersonic, the clusters have thermal velocities.

In general, it is most important to control the temperature of the oven, since in this kind of source only a narrow range of temperature (about 20°C) is available for effective production of clusters. Figure 3.15 shows the cluster mass distributions of Ag beams obtained at different oven temperatures keeping the source pressure at 4 mbar, using a conical nozzle and a growth distance of 35 mm.

The inert gas pressure and the effects of the supersonic expansion are in the expected direction. Up to a few tens of mbar the cluster size increases with the source pressure because of the improved cooling both in the source and in the expansion (Fig 3.16). The reported data for larger variations in the pressure in the source are affected by the detection systems so that no final conclusions can be drawn on the cluster size distribution in this regime.

The growth distance is confirmed to be a key parameter to control the cluster size distribution in the beam: changing the growth distance from 8 to 28 mm can change the maximum of the cluster size distribution by a factor up to 8. The time that the clusters have for growing is crucial to produce larger

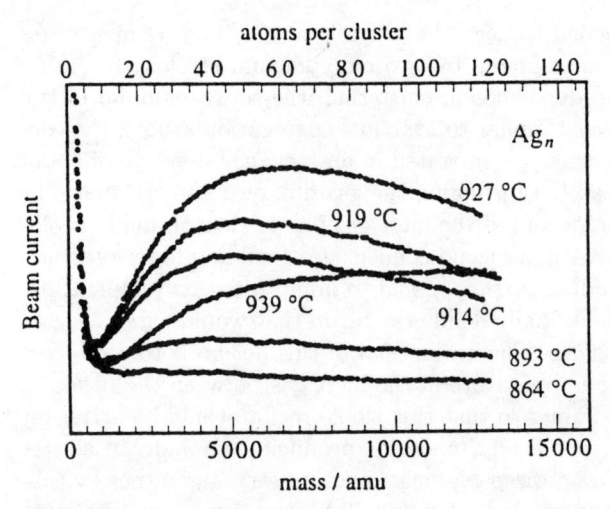

Fig. 3.15. Mass distribution of a beam of Ag clusters for different oven temperatures as obtained from a gas-aggregation source with supersonic expansion. (From [3.61])

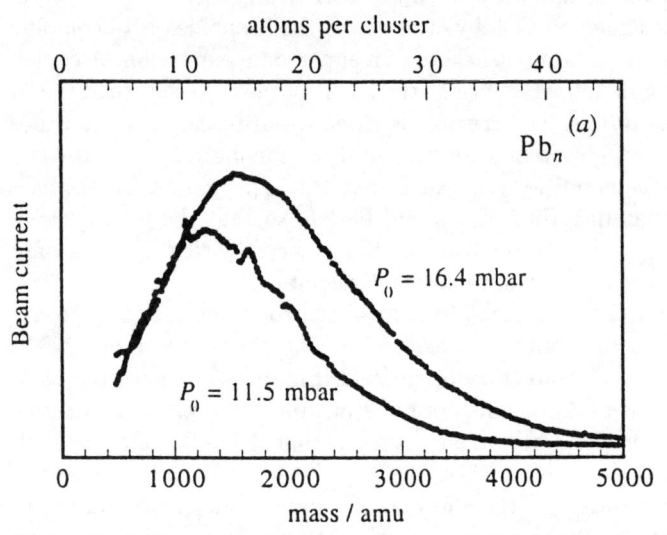

Fig. 3.16. Cluster size distribution in Pb beams prepared by a gas-aggregation source with supersonic expansion at two different inert gas pressures, 11.5 mbar and 16.4 mbar respectively. The growth distance is 8 mm and the conical nozzle is 1 mm diameter. (From [3.61])

clusters. In the case of silver, a growth distance shorter then 10 mm will not give rise to any clustering while distances larger than 28 mm the maximum of the cluster distribution will not move further.

3.2.2 Magnetron Plasma Sources

Sputtering of a target in an inert gas can also be used to form clusters from the aggregation of the ejected neutral atoms and ions. The coupling of a magnetron plasma source to a gas-aggregation cell can efficiently produce nanocrystalline powders [3.63]. Haberland and co-workers [3.64, 3.65] have shown that an effusive cluster beam can be extracted from this kind of source. A schematic diagram of the apparatus is shown in Fig. 3.17.

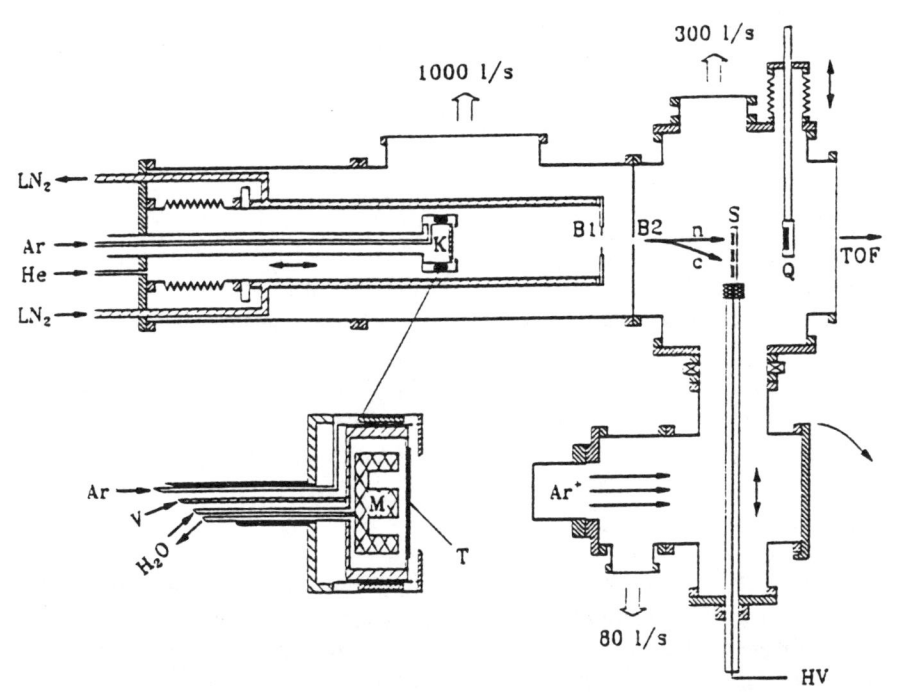

Fig. 3.17. Schematic diagram of a plasma-aggregation source. Atoms are sputtered from the target T of the magnetron discharge head K. The clusters are swept by the gas stream out of the diaphragm B1, pass through B2 and are accelerated onto the substrate S. (From [3.65])

The magnetron discharge is operated at a pressure of about 1 torr to favor aggregation of the sputtered species, clusters are swept though the diaphragm by the gas stream. Ionized clusters can then be accelerated in the second chamber towards the biased substrate. By varying the distance between the magnetron unit and the orifice, the cluster residence time and hence the mass distribution can be controlled. A typical mass spectrum of Cu clusters is shown in Fig. 3.18.

The degree of ionization of the clusters is in general very high, depending on the material (60–80% for Al, 20–60% for Mo, 20–50% for Cu). Deposition

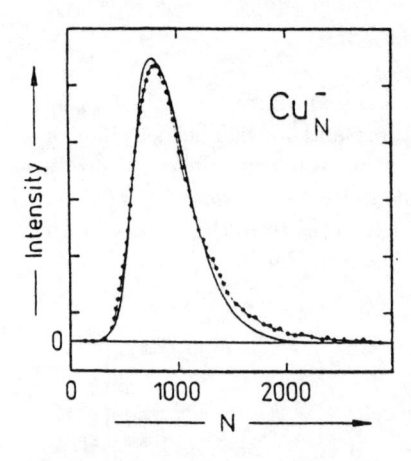

Fig. 3.18. Size distribution of Cu_N^- clusters, obtained with a low-resolution mass spectrometer. (From [3.65])

rates of several hundred Å min^{-1} are obtained with a discharge power of 50–150 W [3.65].

A source based on this principle has been used to grow Cr cluster with good control of the cluster size distribution (Fig. 3.19) [3.66, 3.67].

We can conclude that gas-aggregation sources can be considered as one of the most flexible sources in term of cluster sizes (from 2 to $\geq 10^5$) and in terms of cluster material. Moreover, they offer the important advantage of continuous operation that ensures high deposition rates and small energy spreads.

3.2.3 Supersonic Sources

As discussed in Sect. 2.2, the formation of clusters in supersonic sources arises from the highly non equilibrium cooling experienced by the vapor expanding from a nozzle where the mean free path is much shorter than the hole diameter. During the expansion, the gas density and temperature drop by several orders of magnitude (the cooling rate dT/dt can reach 10^8–10^9 K s^{-1} [3.68]). The cluster growth is regulated by the collision rate during the expansion and will stop when no more collisions occur, so that the final beam composition depends not only on the thermodynamics, but also critically on the kinetic and time scale of the expansion. Therefore, the overall process of clustering, passing from the collision-dominated continuous flow gas dynamics to the collisionless free molecular regime, strongly depends on the initial state (pressure and temperature in the source) and on the nozzle geometry. So far, a realistic description taking into account all the parameters is far from being available, so that only criteria based on the previously discussed scaling laws (see Sect. 2.2.5) can guide the choice of nozzle design.

The nozzle geometry plays a very important role in the design of a supersonic source. Axially symmetric free jets are produced by nozzle geometries whose typical cross-sections are shown in Fig. 3.20.

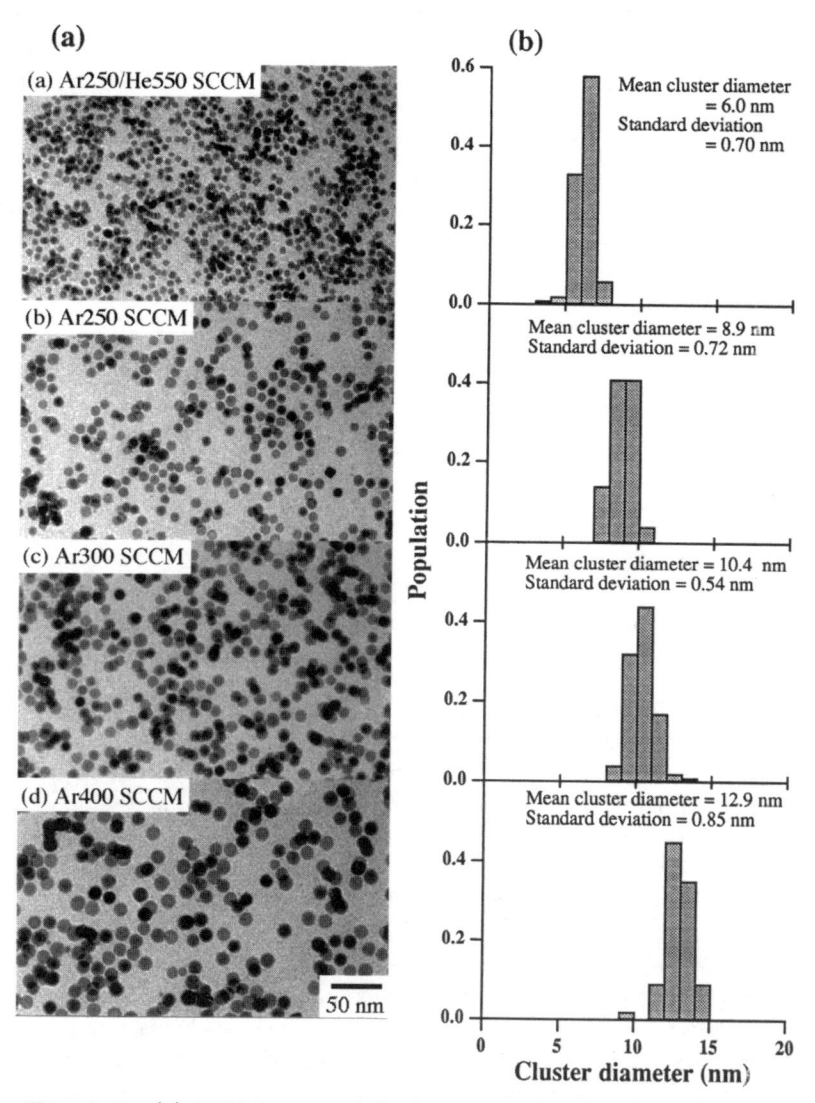

Fig. 3.19. (a) TEM images of Cr clusters produced under different aggregation conditions in a plasma aggregation source, (b) Mean cluster diameter and deposition rate as a function of quenching gas pressure. (From [3.66, 3.67])

The basic sonic nozzle is characterized by a circular, well defined, hole in a wall much thinner than the diameter. Such a design is conceived in order to push the expansion to the limits reducing to the minimum, at comparable stagnation pressures and nozzle diameters, any side effects and, in particular, condensation that in many beam experiments represents a serious problem. The clustering process in this case is minimized by the reduced number of effective collisions (those happening between the onset of supersaturation in

Sonic Nozzle

Conical Nozzle

Laval Nozzle

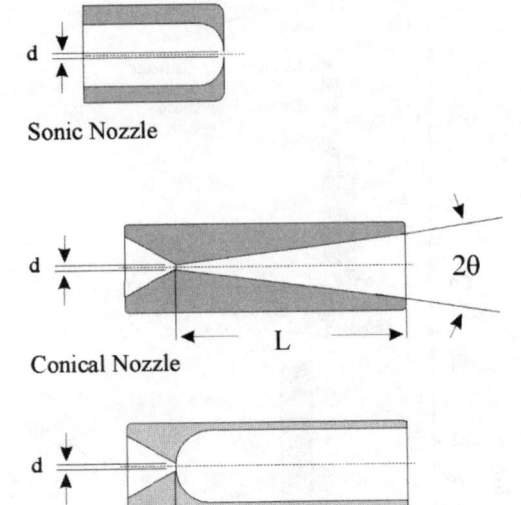

Fig. 3.20. Schematic cross-sections of nozzles with axially symmetric geometries

the expansion and the beginning of the free molecular regime) and the short time during which the expansion occurs. Both of these dramatically limit the cluster growth. One should keep in mind that as the pressure increases, for a nozzle of a given diameter and cross-section and at a fixed source temperature, the supersaturation region moves upstream while the degree of supersaturation grows. Their combined effect results in an increase of the number of useful collisions producing beams with more and larger clusters. As far as the nozzle diameter in axial symmetric beams is concerned, it represents the characteristic length so that, for otherwise fixed source parameters, the beam will have identical conditions at the same number of nozzle diameters downstream. Since, at a given source pressure and for an adequate pumping speed, the terminal beam velocity is independent of the nozzle diameter, smaller nozzles will produce expansions that will run out of collisions sooner, so that smaller and fewer clusters will be present. On the other hand, this same effect will produce much higher cooling rates for smaller nozzles while the effects due to the boundary layer can become significant affecting beam divergence and velocity spread.

Obviously the temperature of the oven has a very strong effect on the cluster size distribution: a colder vapor would favor the onset of supersaturation because of the steep quasi-exponential dependence of the vapor pressure on temperature.

The overall behavior of a nozzle cluster source can be summarized by stating that the cluster abundance and the average cluster size $\langle n \rangle$ in the beam will increase with increasing pressure and nozzle cross-section and with decreasing temperature.

Ideal sonic nozzles are quite difficult to make. In fact, to keep the vacuum system to affordable dimensions, typical nozzle sizes are usually small (a few tenths of a millimeter in diameter or smaller), so that real sources will look more like channels of length comparable to the diameter. Furthermore, several methods used for the production of the hole cannot really maintain a sharp edge, so that the hole usually has a conical shape converging (or diverging) towards the outside of the source (converging–diverging nozzles). These different cross-section shapes can have severe effects on the overall performance of the nozzle, changing the angular distribution of the beam, the cooling rate and the cluster size distribution. In particular, a constrained expansion may produce larger clusters because the vapor experiences more collisions before the expansion ceases. The basic idea of using these effects to constrain the expansion in a controlled way has been pioneered by Becker et al. [3.69], Hagena and Obert [3.70] and developed by Stein and Armstrong [3.71], Hagena [3.52] and Gspann [3.72]. A mathematical modeling of flow, cooling and clustering in these shaped nozzles is not yet fully achieved so that the design of the nozzle shapes is still complex and empirical, strongly limiting a widespread use of them in different potential applications.

Simple conical, cylindrical and converging-diverging nozzles have been reliably used in the production of cluster beams and their performances have been compared. For a conical nozzle it has been shown that the beam profile that is obtained follows roughly the cone shape. Nozzles with narrower cone angles produce beams with very high forward intensities [3.73] and one should take great care of the effects due to the boundary layer and to skimmer interference. Even though electroformed skimmers guarantee razor sharp edges, the interference due to the scattered particles seems to be the limiting element to improved intensities. Deposits on the skimmer due to the same beam material could drastically reduce the intensity in a short time.

If one compares the performance with a standard circular sonic nozzle under similar throughput conditions, the nozzle with conical cross-section gives rise to beam intensity improvements of at least one order of magnitude. Important consequences are then that a larger number of binary collisions will be present (responsible for cooling in the supersonic expansion) and of three body collisions (required for the cluster growth). Another way of looking at the properties of conical nozzles is to say that the pumping system requirements to produce the same cluster size and density in the beam will be much less stringent. The flow on the jet axis, which guarantees the conditions for clustering, remain unchanged for a conical nozzle where the nozzle contour matches the free-jet streamtube determined by the cone angle 2θ. On the other hand, the total flow passing through this conical nozzle is much smaller compared to the equivalent sonic nozzle. This can be schematized by the concept of the equivalent nozzles [3.52, 3.53], that states that the properties of a conical nozzle scales to that of a sonic nozzle by defining an equivalent nozzle diameter d_eq given by:

$$\frac{d_\mathrm{eq}}{d} = \frac{0.74}{\tan\theta}. \qquad (3.3)$$

To give an idea of how consistent the effect on the gas load and hence on the vacuum system can be, let us consider a cone angle $2\theta = 8°$. In this case the ratio $d_\mathrm{eq}/d = 10.6$ so that the mass flow reduction would be $(10.6)^2$: more then two orders of magnitude.

This is only an idealized picture: in particular, at small cone angles the effect of the boundary layer could seriously affect the expansion (reducing the effective θ) and eventually destroy the supersonic beam. Furthermore, the condensation process, heating the beam, widens the free-jet streamtubes [3.74]. The use of (3.3) gives, therefore, only an estimate of the beam properties while the effective performance could be slightly better [3.70] or worse [3.53]

Since the pioneering work by Becker et al. [3.69] inert gases have been added to the vapor to promote the formation of clusters. The mass of the inert gas and the dilution rate (seeding) then become two other important parameters to adjust to control the beam mass distribution. During the expansion of the vapor seeded in the inert gas, the carrier gas atoms will play the very important role of cooling the growing cluster by taking away the heat of condensation by collisions. This produces a strong effect of stabilization of large clusters. If the dilution is obtained by increasing the pressure of the carrier gas, the process is very efficient and a saturation effect is observed only at very high dilutions. This is the case of joule heating vaporization of metals where usually the vapor pressures are low. Of course, for gaseous materials the beam composition can be varied by changing, independently, the pressure of the two gases in the mixture so that, for a fixed pumping system, a trade-off should be sought in order to reduce the effects of an excessive dilution that would reduce the probability of collision between the atoms of the seeded species.

Figure 3.21 shows the effect of different noble gas atoms, used as carrier gas, on the size distribution of Na clusters formed by a cylindrical nozzle at a fixed source temperature and at the same pressure.

The behavior observed in Fig. 3.21 for the alkali metals is quite general with a strong increase of cluster size going from the lighter to the heavier carrier gas: heavier carrier produces a slower expansion so that clusters have more time to grow. This is, of course, limited to the case of large differences in condensation energies between the carrier and the seed. Otherwise the formation of solvation complexes readily occurs changing the final size distribution of clusters in the beam in a critical way up to showing the opposite behavior. This is the case of SF_6 seeded beams where the average cluster size decreases when heavier noble gases are used as carriers [3.76].

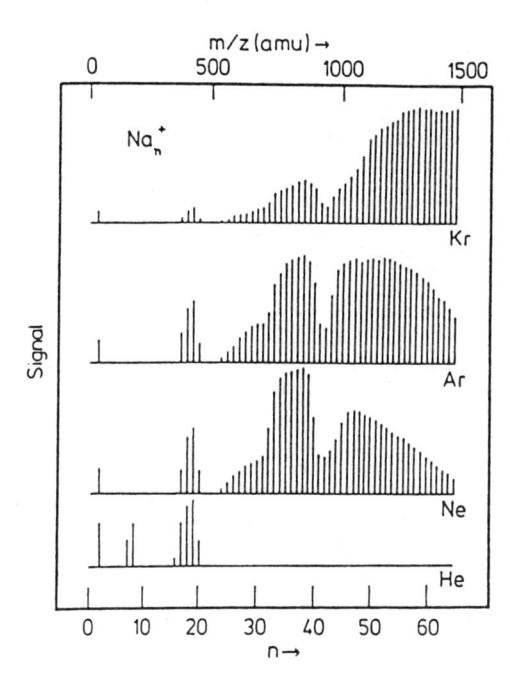

Fig. 3.21. Mass spectra of beams of sodium seeded in inert gases. The same cylindrical nozzle of 0.3 mm diameter, at a temperature of 1070 K (Na vapor pressure of 350 torr) and inert gas pressure of 1.3 bar is used for all the photo-ionization spectra. (From [3.75])

When the vapors are molecular another important effect of the seeding occurs: a monatomic carrier gas induces a much more effective cooling of the molecules. Because of this gas-dynamic effect, the production of clusters of polyatomic molecules critically depends on the adoption of the seeded beam technique [3.77]. On the other hand, when it is very difficult to produce a sufficiently high vapor pressure of a material, the use of a carrier gas becomes the only practical way to give rise to a supersonic beam of clusters.

Another aspect that is very relevant in thin film growth and surface modification processes, even though not fully exploited up to know, is the effect of gas-dynamic acceleration due to the supersonic expansion.

As discussed in Sect. 2.1.2, the final flow velocity in an isentropic expansion is given by:

$$\overline{v_\infty} = 1.581\sqrt{\frac{2kT_0}{m}} = 204\sqrt{\frac{T_0}{m}}, \qquad (3.4)$$

where, in the last term of the above equation the mass is expressed in atomic mass units. A fast-flowing light carrier gas will therefore produce a strong gas-dynamic acceleration on the very diluted quantity of slow-moving clusters. Depending on the mass of the cluster and on the temperature of the source, energies of several tens of eV can be easily reached [3.78–3.81]. Assuming a single cluster species of mass M_c seeded in a carrier gas of mass M_g the maximum kinetic energy achievable will be:

$$E_k = \left(\frac{\gamma R T_0}{\gamma - 1}\right)\frac{M_c}{M_a}, \qquad (3.5)$$

where M_a is the average molecular weight, which reduces to M_g in the limit of high dilutions. The highest gas-dynamic acceleration would then be observed for hydrogen mixtures because of its higher $\gamma/(\gamma-1)$ due to rotational relaxation effects. In terms of energy per atom, depending on the mass of the clusters and on the temperature at which the source is operated, values of a few eV can be easily reached (roughly 5–10 times larger than for an evaporative source).

A typical supersonic source setup is shown in Fig. 3.22. It has been developed by the Karlsruhe group to prepare cluster beams of different high melting point materials such as Ag, Mg, MgF$_2$ and PbF$_2$ [3.78].

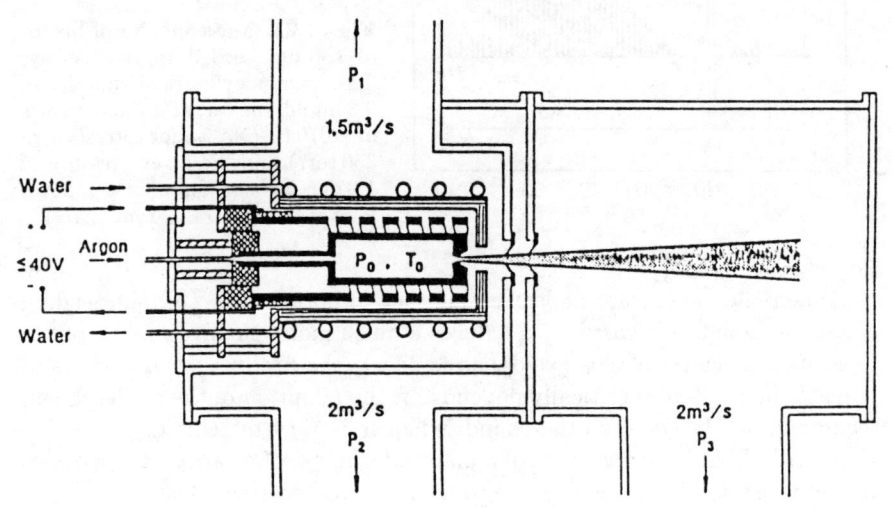

Fig. 3.22. Schematic of the cluster beam system based on a seedable conical supersonic nozzle developed by the Karlsruhe group. (From [3.82])

As is typical of supersonic beams, the vacuum systems should be adequate for the required gas-load. In the set up of Fig. 3.22 a two-stage differential pumping is used before the deposition vacuum chamber. The pumping in the source chamber is obtained by a refrigerator cryopump designed for gas-loads up to 28 mbar l s^{-1} corresponding to 23 h of continuous operation. The other differentially pumped stages are evacuated by 2000 l s^{-1} turbomolecular pumps.

The cylindrical nozzle assembly (about 42 mm diameter) consists of a gas supply line, an oven of about 50 cm^3 and a screwed on nozzle, all of which are made of high-density graphite so that it can be heated up to 2350 K. The heater, made of graphite, is a double-wound coil (square cross-section 12 × 4 mm^2) machined out of a cylinder 58 mm diameter that is supplied by a power of 3 kW to reach maximum oven temperature. A multiple tantalum shielding system carefully insulates the oven in order to keep the power requirements to a minimum. The nozzle is conical with diameter ranging from 0.35 to 1 mm and sections with apex of 10° and 17° and length of 17 mm. The skimmer is heatable to temperatures well above the melting point of the vaporizing material (1300 K for Ag that has a melting point of 1234 K). Different diameters are used ranging from 1.4 to 3 mm at distances from the nozzle of 36 and 102 mm respectively. This type of source, using Ar as carrier gas at pressures up to 5 bar, produces beams of cluster of Ag with sizes ranging from 50 to 1000 atoms per cluster and an average kinetic energy of about 1 eV per atom at 1500 K. With an appropriate choice of skimmer and collimators the profile of the beam at about 1 m from the source is quite uniform over an area of 120 mm^2, where a deposition rate of about 10 nm s^{-1} is achieved. The use of a conical nozzle also produces the great advantage of reducing to a minimum the waste with 20% of the material originally evaporated in the nozzle going into the beam flux, and therefore being deposited [3.78].

Sources for bimetallic clusters have also been developed. Here, complications arise from the need to control independently the vapor pressure of the two species. This goal becomes particularly difficult to achieve when materials with very different melting points such as alkali and alkaline earth or even transition metals are considered. In this case, stable temperature differences between the two species have to be maintained on a long time scale to have reproducible composition of the beam. This problem has been experimentally resolved by the Bern group [3.83] by coupling two shielded ovens, with the one at lower temperature being upstream. The two cartridges are insulated one from the other by a 5 mm thick molybdenum plate and a 5 mm thick layer of binder free ceramic. This insulation guarantees a temperature gradient of about 1000 K over a few hours of operation. Na$_x$Au$_y$ and Na$_x$Ag$_y$ clusters are efficiently produced by this source, although containing only a small number of clusters. Even though such a solution is very useful to study bi-metallic clustering the system is quite complicated both to build and to operate so that it is not very practical for thin film deposition.

For materials that can be efficiently vaporized at lower temperatures an efficient and practical design can be obtained using quartz tubes. Such a solution has been adopted by Nagaya et al. [3.84] using a quartz tube ending with a converging nozzle of about 0.1 mm diameter. The reservoir is also made of quartz and connected to the side of the main quartz tube. The source has been operated at 670 K to produce clusters of selenium by seeding it in Ar at 10 torr. Figure 3.23 shows a different design developed by the Trento group [3.80, 3.81] to produce supersonic seeded beams of fullerene, of organic molecules and of oligopolymers.

It basically consists of two concentric quartz tubes. Cylindric and conical nozzles are formed by hot squeezing a capillary to the desired shape. The nozzle is then attached to the two-tube assembly by careful melting, while rotating the whole system on a lathe. This is the critical part in the construction of this kind of source since it is not easy to preserve the shape

70   3. Cluster Sources

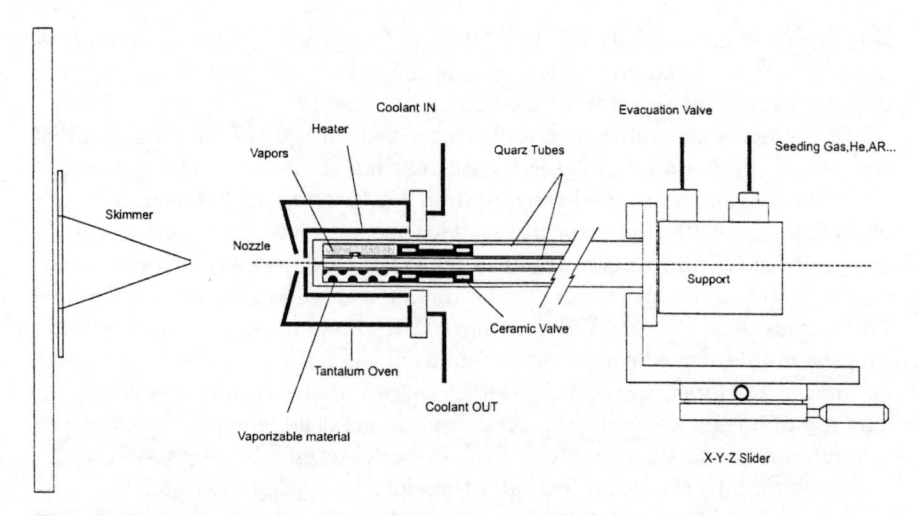

Fig. 3.23. Schematic view of the Trento supersonic source made of two concentric quartz tubes. The space between the two tubes is used as a reservoir. The vapors flow into the inner one where they seed the flowing carrier gas

of the nozzle during the melt-welding process. The volume between the two quartz tubes is the reservoir where the material to be vaporized is inserted. A movable sector made of machinable ceramics (both Macor and Shapal M have been used) allows to insert the material and to isolate it from the back of the source. The reservoir volume is typically a few cm^3 and is heated by tantalum wires or a tantalum strip in close contact with the outside of the quartz tube. The front part of the nozzle is heated to slightly higher temperatures by a separate element. A series of tantalum shields isolate the heater. A small, pressure-operated valve (bellows sealed) allows evacuation and outgassing of the system. This compact and efficient design allows highly supersonic expansion with carrier gas pressures up to 4 bar with a nozzle diameter of 0.075 mm and a pumping speed in the source chamber of $2000\,\mathrm{l\,s^{-1}}$. It operates at temperatures higher than 1700 K.

3.3 Pulsed Sources

3.3.1 Pulsed Valves

Pulsed valves are widely used for the production of cluster beams because of their low duty factor. They facilitate strong expansion and high instantaneous intensities with a moderate pumping speed requirements. Moreover pulsed beams can be coupled very efficiently with pulsed spectroscopic and detection systems such as pulsed lasers and time-of-flight mass spectrometers [3.85] (see Chap. 4). On the other hand the low duty cycle represents an obstacle to the use of pulsed cluster beams for the synthesis of nanocrystalline materials, since the attainable deposition rates are very low, especially if compared to continuous sources. This is the basic reason why the use of pulsed cluster sources is often confined to the characterization of clusters in beams. Improvements in the design and operation of pulsed cluster sources have stimulated their use in the production of nanostructured materials.

Pulsed sources consist of an oven, where the material is vaporized, coupled with a mechanical device (pulsed valve) capable of delivering a pulse of gas with suitable characteristics of duration and intensity. The basics of pulsed valve design and a discussion of the mechanical limitations to take into account are reported in detail in [3.85], here we will present several pulsed valve designs and discuss their relevance for the design of a pulsed cluster source.

One of the most popular pulsed valves for the production of clusters is the solenoid valve. It is actuated by passing a current pulse through a solenoid, which exerts a magnetic force on a ferromagnetic core material, usually iron. The prototype of this family of valves has been realized by adapting a commercial automobile fuel injector [3.86], as schematically shown in Fig. 3.24.

A fuel injector can be modified by the introduction of a stainless steel nozzle to replace the original valve seat, the main body of the fuel injector is press fit into a stainless steel collar that permits adjustment of its axial position with respect to the plunger. The plunger is removable and it can be

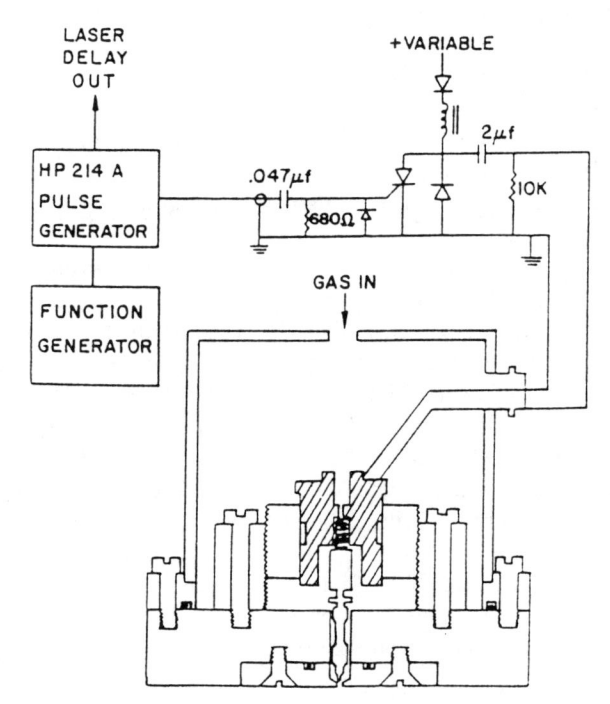

Fig. 3.24. Schematic representation of the fuel injector pulsed valve. (From [3.86])

lapped to assure a tight seal. The valve assembly is encased in a pressure can which can hold high gas pressures.

The temporal shape of a gas pulse emerging from this type of valve has been characterized in a multiphoton ionization experiments with a NO–He seeded beam. The beam was skimmed approximately 3 cm downstream of the nozzle by a 1.33 mm skimmer and intersected by the laser light approximately 8 cm downstream of the nozzle. The source produces a gas pulse 600 μs wide, consisting in a peak with a full width at half maximum of ∼ 150 μs and a long tail. Due to these characteristics, and depending on the backing pressure, pumping speed and position of the skimmer, a shock wave can form at some point in the pulse resulting in a dynamic shutter effect which limits the gas flow through the skimmer [3.87]. The source was operated at pulse rates up to 50 Hz.

A more sophisticated solenoid valve is described in [3.88]. The actuating mechanism uses two solenoids, one to open and one to close the valve and it is isolated from the source gas chamber by a small bellow. A Viton-tip plunger is pressed against the nozzle by a small holding current circulating in the front solenoid. A removable front plate consists of a stainless steel disk with a conical hole bored in its center from one side. On the other side of the disk, concentric with this hole, sits an electron-beam-welded 800 μm nickel aperture. This acts both as nozzle and as valve seat [3.88]. To open the valve, the holding current is turned off, a larger current pulse is provided to the back solenoid. At the end of the opening pulse, the back coil is switched off and the forward one turned on. The open pulse width and amplitude can be adjusted to provide the desired pulse duration. A 70 μs FWHM pulse represents the minimum width reported (Fig. 3.25).

Under typical operating conditions, an intensity of 10^{22} molecules $sr^{-1}s^{-1}$ is estimated. Since power dissipation in the coils is very low, operation up to 35 Hz with a very modest power supply is possible [3.85].

A scheme of a compact solenoid valve commercially available [3.89] is shown in Fig. 3.26.

The actuating mechanism is based on one solenoid, the plunger protrudes into the solenoid and it consists of a metallic armature holding, at the front end, a teflon poppet with a conical tip. The teflon tip is maintained in the closed position by a spring inserted in the armature, which in turn is maintained in place by another buffer spring. The frontplate part of the valve containing the nozzle is provided with a screw thread and can be dismounted from the main body of the valve. The position of this piece is critical in determining the position of the plunger in the solenoid and the working conditions of the two springs. These parameters are critical for the opening of the valve, for a tight sealing, and for the teflon tip consumption rate. The performance of this valve has been characterized in [3.90]: a typical gas pulse temporal evolution is shown in Fig. 3.27. The valve can operate at pulse widths down to 160 μs, with backpressures up to 85 atm. at 120 Hz.

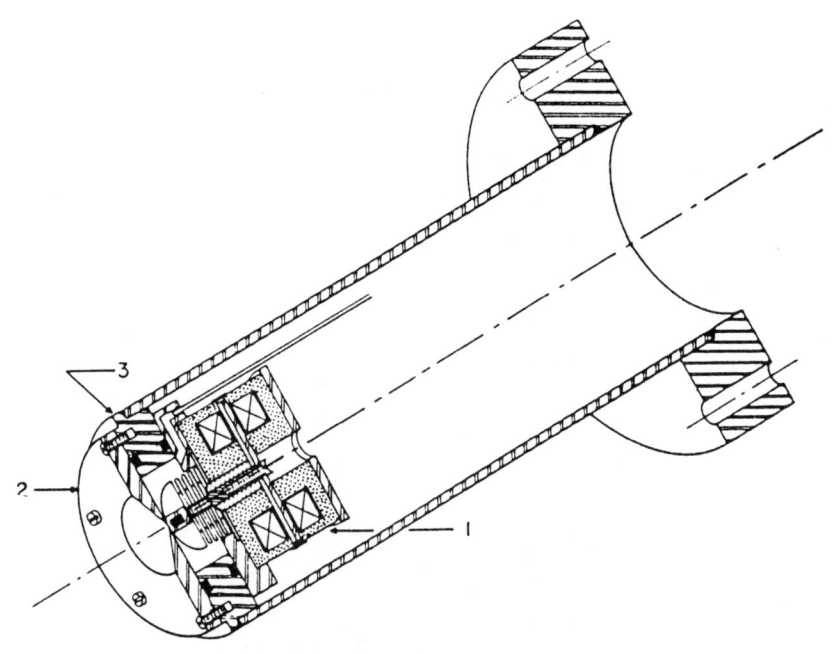

Fig. 3.25. Cutaway view of a solenoid pulsed valve: 1 actuator assembly, 2 end plate with nozzle, 3 mounting ring with extension tube and back flange. (From [3.88])

A valve based on a similar principle, where an aluminum valve stem placed between an opening and a closing coil, is pressed on a viton o-ring behind the nozzle by a leaf spring carbon fiber, is described in ref. [3.91]. This valve can deliver pulses with intensity of 10^{22} molecules $sr^{-1}s^{-1}$ (operating with He) with opening times variable between 10 and 100 μs.

Another class of pulsed valves are the piezoelectric ones. They have an actuating mechanism based on a flexing disk of piezo material attached to a metallic membrane, the high resonance frequency of the disk ($\sim$ kHz) and low power consumption allow operation at very high repetition rates. The model most widely used and also commercially available is based on a modified version of a leak valve [3.92, 3.93] (Fig. 3.28). Pulses of duration as short as 100 μs or as long as several milliseconds can be obtained at repetition rates as high as 750 Hz. This valve has the disadvantage of a small excursion of the piezo disk which results in a nonlinear dependence of the total gas throughput from the backing pressure.

This drawback can be partially circumvented by simply increasing the driving voltage on the crystal or by loosening the mounting screws of the piezo disk. However this can result in multiple pulsing due to the reduced damping of the membrane [3.94, 3.95]. Proch and Trickl [3.95] have proposed a piezoelectric valve characterized by a much stronger disk translator compared to the original design. This results in a large excursion of the piezo disk and in a high mechanical stability and long-term reproducibility of the valve

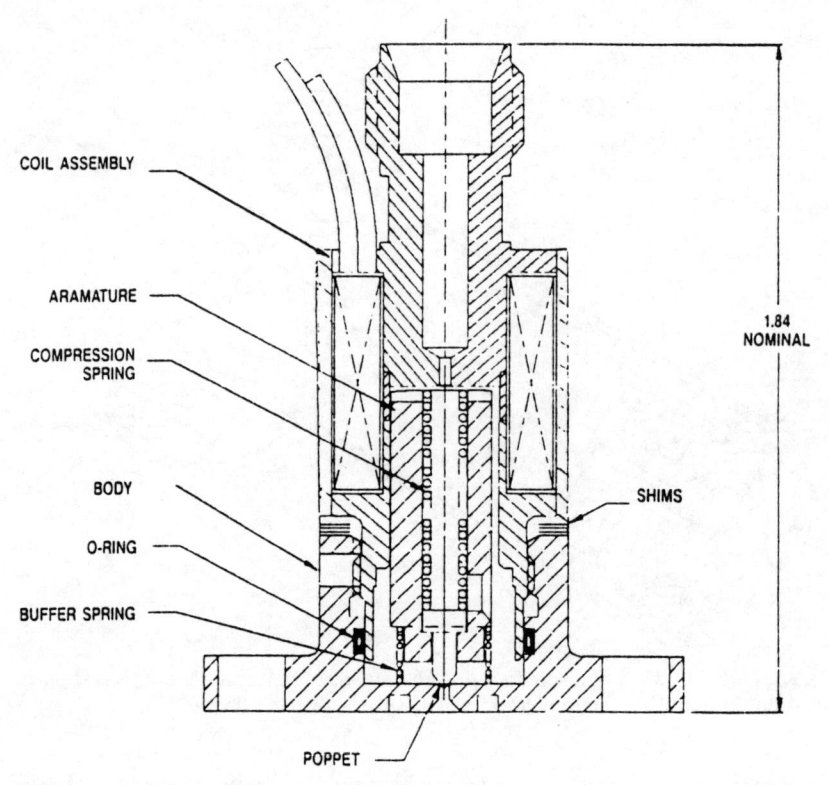

Fig. 3.26. General Valve PS-9. (From [3.89])

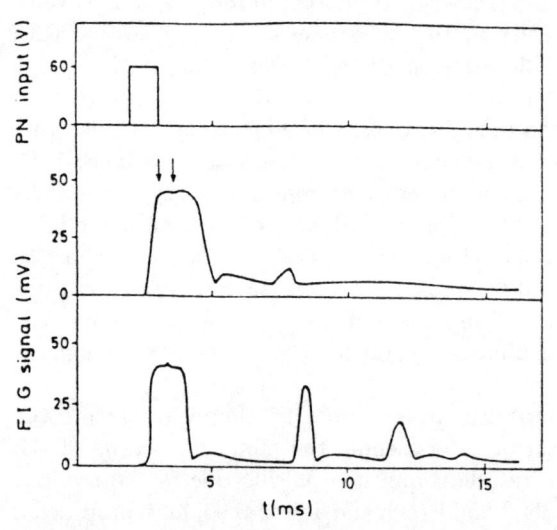

Fig. 3.27. Performance of a General Valve PS-9: (a) voltage driving pulse, (b) Fast Ion Gauge characterization of the gas pulse shape delivered by the valve. (From [3.90])

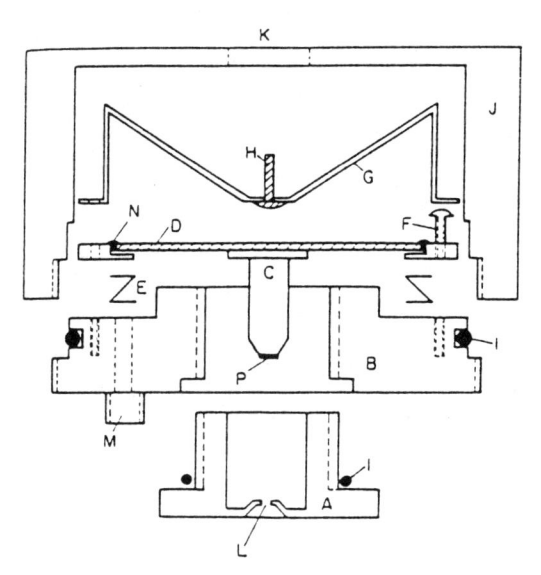

Fig. 3.28. Modified piezoelectric valve: A–nozzle insert, B–valve face plate, C–plunger, D–piezoelectric crystal, E–springs, G–springs, F–spring tension adjustment screw, H–insulating rod, I–o-ring seal, J–valve body, K–gas inlet, L–nozzle orifice, M–electrical feedthrough, N–silicone rubber seal, P–viton seal. (From [3.93])

performance. Another improvement is an adjustable plunger allowing easy adjustments and better sealing. The maximum possible flow can be achieved for voltage pulse widths larger than 150 μs which results in gas pulse widths ranging form 170 to 250 μs, depending on the selected nozzle.

Very short gas pulses ($\sim$ 10 μs) can be produced by current-loop actuated valves [3.85, 3.96] whose operating principle is schematically shown in Fig. 3.29. The seal is made by a metal bar the center of which rests on the o-ring surrounding the high pressure side of the nozzle. Nozzle diameters up to 0.75 mm can be used with a good sealing. The actuating force comes from a current pulse circulated through a loop formed by the faceplate and the metal bar. The magnetic force generated by the opposing currents pushes the bar away from the o-ring, allowing the gas to flow through the nozzle. In order to obtain short pulses, huge driving forces must be applied and hence a fast power supply must be available [3.85].

All the pulsed sources described up to now suffer the limit of a narrow range in operating temperatures. Due to the presence of mechanical parts and rubber sealing, they can be safely operated only at temperatures close to room temperature. The use of pulsed sources at very high or low temperatures can be realized by isolating the heated part from the actuating mechanism or by using materials whose dimensions and mechanical properties are constant over a wide temperature range. A solenoid valve that can operate up to 550 °C has been designed by Li and Lubman [3.97] (Fig. 3.30). Due to the mass of

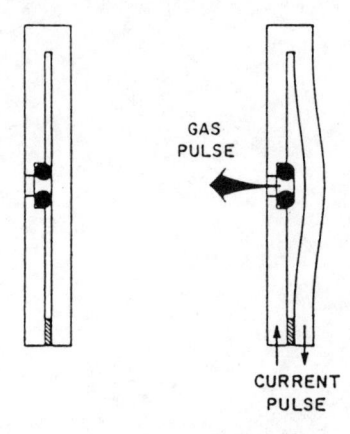

Fig. 3.29. Principle of operation of a closed-loop valve: with application of a current pulse, the repulsive force flexes the bar and allows gas to flow through the o-ring. (From [3.96])

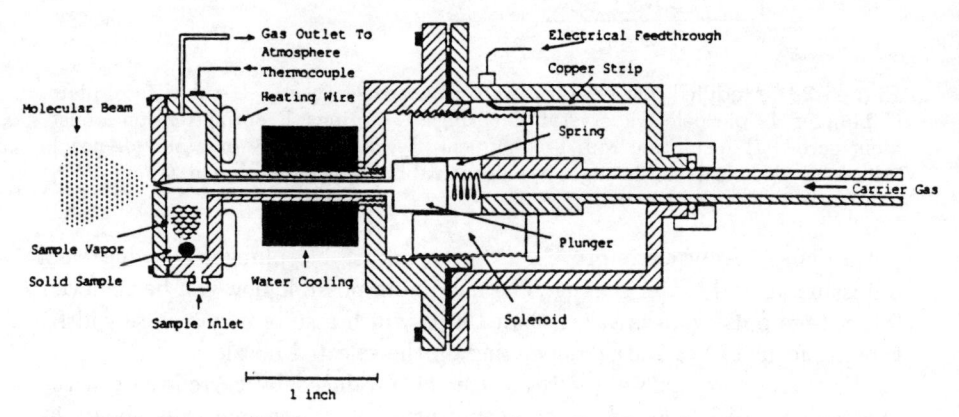

Fig. 3.30. Design of high-temperature pulsed nozzle. (From [3.97])

the plunger, high intensity current must be provided to the solenoid. Gas pulses of $\sim 225\,\mu s$ FWHM are reported for organic molecules seeded in Ar. The pulses reach a flat top with $\sim 130\,\mu s$ width indicating that choked flow is reached. A similar constructing solution could be adopted also with the piezoelectric valve described in [3.95] and for the General Valve model [3.98].

Pulsed valves or experiments at cryogenic temperatures are proposed in [3.99, 3.100]. Bucher et al. [3.99] have modified a General Valve 9 by permanently bonding the two threaded pieces of the valve and adding a new access to the poppet which seals with an indium wire. In this configuration, it can be supplied with He cooled down to liquid nitrogen temperature.

Hagena [3.100] also adopts a solenoid mechanism with a sealing poppet made of Kel-F that provides a perfect seal down to 20 K.

3.3.2 Laser Vaporization Sources

The coupling of laser vaporization to supersonic molecular beams sources for the production of clusters was proposed in 1981 by Smalley and co-workers [3.101]. The design and realization of LVCS are very simple, especially if compared to hot oven sources. LVCS are similar to the chambers used for laser ablation for thin film deposition, the only relevant difference is that the plasma plume expands in a buffer gas.

A schematic representation of a LVCS is shown in Fig. 3.31. The light of a high intensity pulsed laser (usually a Nd:YAG laser operating at 532 nm, pulse width 5–10 ns) is focused onto a target rod and a small amount of material is vaporized into a flow of an inert carrier gas. The inert gas quenches the plasma and cluster condensation is promoted. The mixture is then expanded in vacuum and forms a supersonic beam [3.23]. LVCS are operated in a pulsed regime hence pulsed valves are used for delivering the carrier gas. This allows the use of a relatively economical apparatus with moderate pumping speed.

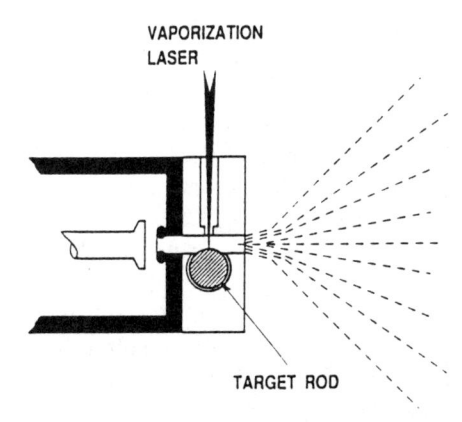

Fig. 3.31. Schematic diagram of the principle of operation of a standard LVCS. (From [3.23])

Several factors affect the LVCS performance in terms of intensity, stability and cluster mass range attainable: quantity of ablated material, plasma–buffer gas interaction, plasma–source wall interaction, and cluster residence time prior to expansion. As we will discuss in more detail below, a careful choice of the source geometry allows an optimization of the relevant parameters and a substantial improvement of the source stability and intensity.

In a standard source, the inert gas flow is directed through a narrow channel (typically 2 mm diameter). Perpendicular to this channel a hole is drilled to allow the light to illuminate the target rod. The laser pulse vaporizes a small amount of the material which is ejected as a plasma extending up to 10 mm above the surface of the rod. Although most of the energy in the plume is transferred to the walls of the channel by thermal conductivity through the inert gas, a reasonable amount remains in the gas, generating relatively hot beams with high mean velocities. Better thermalization is usually achieved by using long channels (2–3 cm) or using some pre-expansion extensions of larger diameter [3.102, 3.103]. However, a long channel length allows the metal vapor and clusters to diffuse to the walls and hence be lost. Another disadvantage of this configuration is that the material ejected in the plume deposits on the wall of the channel and on the hole through which the laser radiation enters. The interaction of the radiation with these deposits produces an opaque plasma that shades the intensity of the radiation reaching the rod and thereby causes unstable operation.

To overcome these shortcomings, a cavity can be introduced where the vaporization takes place before the channel. This cavity should be properly dimensioned in order to minimize the interaction of the plume with the walls, thereby reducing material deposition, and optimizing heat transfer to the cavity walls by the inert gas. Furthermore, the processes involved in the vaporization, cluster formation and thermalization are effectively decoupled from the expansion process [3.24].

A horizontal cross-section of a source with a thermalization cavity is shown in Fig. 3.32. The source body is machined out of a rectangular stainless steel block. Focused laser light enters the source via the 4 mm diameter, 40 mm long channel that is also used to inject the carrier gas onto the target. A plug with a smaller diameter hole (2 mm) is inserted at the entrance of the light channel to reduce the gas losses. The channel terminates in a horizontal cylindrical cavity of 8 mm diameter capped with an adjustable teflon plug on the rear and a teflon nozzle on the front. As shown in Fig. 3.32, the target rod is inside a thin-walled vertically mounted stainless steel tube which intersects the vaporization cavity. A 3 mm hole, coaxial with the light channel is drilled through this tube. When the metal rod is withdrawn, the laser light can pass through the source, so that an accurate alignment can be obtained. The metal rod is pressed against the tube by a spring and a 2.5 mm steel ball (Fig. 3.32) in order to minimize the gas leaking past the rod. This is very important for a stable operation of the source as well.

In Fig. 3.33 we show the source in the vacuum chamber. The source is coupled directly to the pulsed valve by removing the original valve front plate [3.88, 3.89]. The valve is mounted on a bellows which allows external adjustments of the source, relative to the skimmer, in three dimensions. The laser light enters the vacuum chamber via a window. A second window diametrically opposed to the first allows the laser to be aligned with the source, as described above. On top of the source is positioned the rod rotation/translation assembly, which consists essentially of a fine thread screw to which the rod is fixed. During operation the screw is slowly rotated to always present a fresh surface to the laser pulse and to prevent the drilling of a hole in the rod.

During standard operation, with backing pressure of several bars, the maximum pressure in the thermalization cavity reaches about 0.5 bar. The geometry of the source does not favor a high supersonic expansion since the

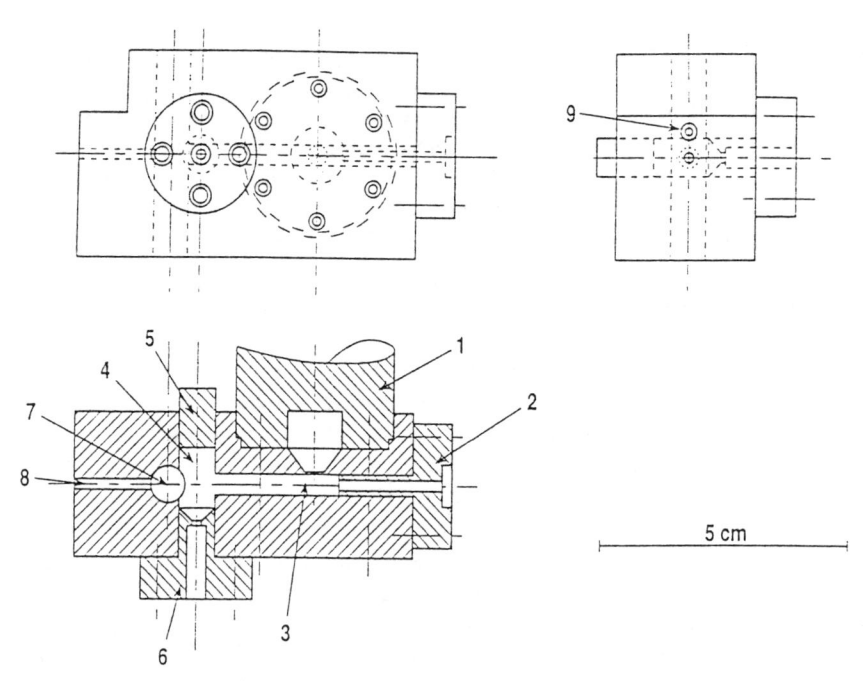

Fig. 3.32. Horizontal cross-section and side view of a LVCS with thermalization cavity: 1–pulsed valve head, 2–metallic plug to reduce gas losses, 3–light channel, 4–vaporization cavity, 5–adjustable teflon plug, 6–nozzle, 7–rod housing, 8–alignment tube coaxial with the light channel, 9–housing for a spring and steel ball. (From [3.24])

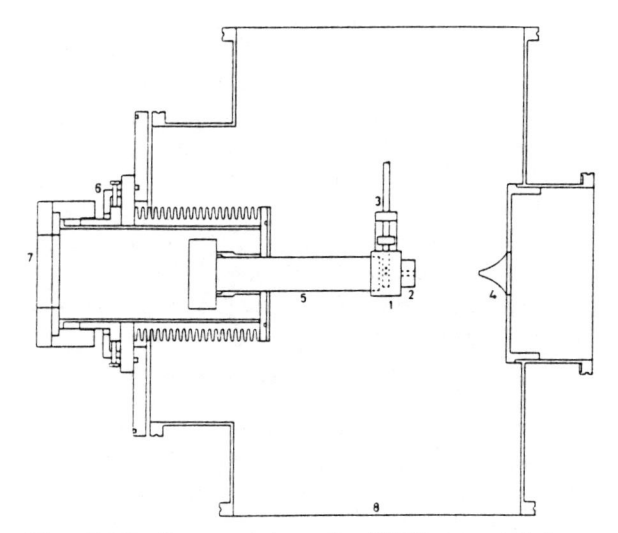

Fig. 3.33. Cutaway view of a LVCS mounted in a vacuum chamber: 1–LVCS, 2–nozzle, 3–rotation/translation assembly, 4–skimmer, 5–pulsed valve, 6–external adjustments mounting, 7–gas entrance. (From [3.24])

gas channel is not straight, however, this configuration optimizes the confinement of the ejected plasma, its mixing with the carrier gas, and thermalization. The characteristics of the cluster population are controlled by the local gas pressure during plasma production and residence time of the particles in the source body. The plasma gas interaction affects not only the final cluster distribution but also the subsequent expansion and beam formation. By monitoring the pressure evolution in a LVCS [3.104], it has been shown that vaporization in a low pressure environment produces a large amount of monomers. Increasing the pressure during the ablation results in a shift of the cluster distribution towards larger masses (Fig. 3.34).

Mass spectra of aluminum clusters are shown in Fig. 3.35 as an example of the source performances. Each peak of the single-shot mass spectrum is produced by about one hundred ions. The signal is stable for about one million shots. Positive ionized clusters are detected without ion optics.

Cluster velocities depend primarily on the time delay between vaporization and detection. Figure 3.36 reports the results for two measurements taken by varying the delay time Δt between the laser pulse and the trigger of the detector. Comparing the flight times determined from the measured velocities with Δt, the dwell time of particles in the source can be determined and are found to be typically on the order of 100 ms. This may be compared with typical dwell times in sources without the thermalization cavity which are estimated to be about 10 ms.

The use of a thermalization cavity allows the separation of the processes of vaporization, mixing-thermalization and expansion. The large amount of heat in the vaporization process is carried away by the inert gas to the cavity walls and does not significantly affect the following expansion. Moreover the dimensions of the cavity are chosen in order to optimize the cooling of the plasma plume and minimize the loss of vaporized material to the walls. Changes in the cluster size distribution are observed by changing the mixing chamber volume. The evidence that cluster formation occurs in the cavity is that, unlike the standard source, the length of the nozzle channel does not significantly affect cluster production. The most critical parameters for cluster production are the alignment and focusing of the laser spot on the target and the proper synchronization of the gas and the vaporization laser pulses.

It should be remembered that the presence of a cavity produces very long gas pulses exiting from the source. This can favor the formation of shock waves in the source chamber depending on the backing pressure and on the nozzle–skimmer distance.

As already discussed in Sect. 3.1.3, laser pulse characteristics influence the quantity and the state of the matter removed from the target. Pellarin et al. have shown that the use of a Ti:sapphire laser (790 nm, 30 ns pulse width) can produce more intense transition metal cluster beams over a wider mass range [3.105]. Deposition rates of $0.2\,\mathrm{nm\,s^{-1}}$ for transition metals have been reported. Although the microscopic mechanisms underlying this improvement

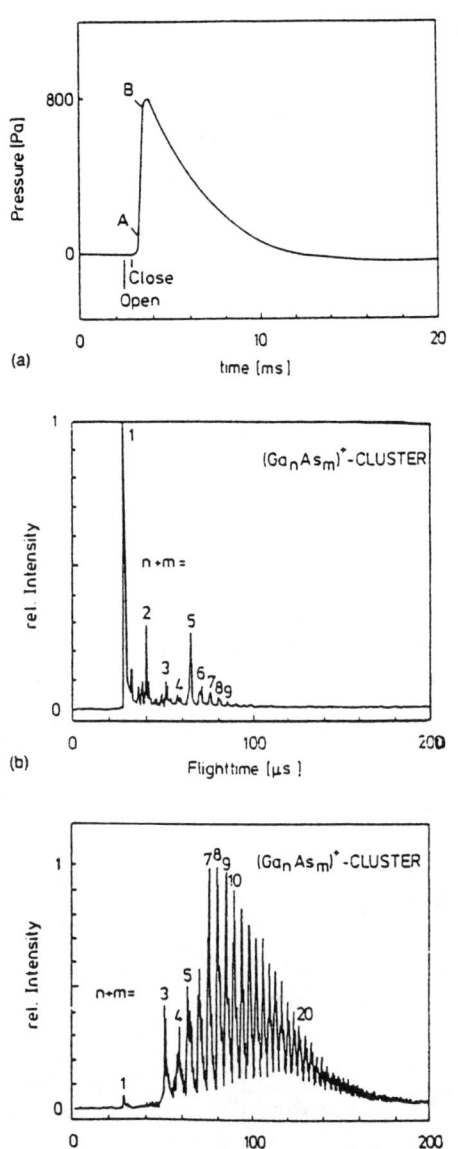

Fig. 3.34. (a) Pressure curve for a valve opening time of of 325 μs, (b) TOF spectrum of Ga_nAs_m produced by firing the vaporization laser at the beginning of the filling of the vaporization cavity, (c) mass spectrum corresponding to firing at the establishment of the maximum pressure in the cavity. (From [3.104])

have not been characterized in detail, the longer pulse duration should be responsible for a more efficient material removal from the target [3.33].

Different geometries have been used depending on the target shape: when the target is a disk (as for many semiconductor materials), a gear mechanism must be used in order to allow a uniform consumption of the surface of the target. In particular, a hypocycloidal planetary gear assembly has been

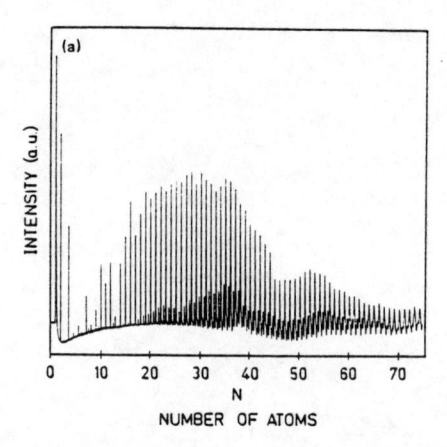

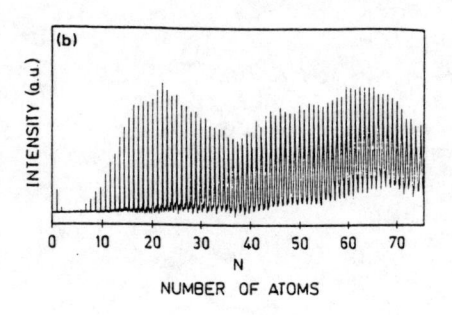

Fig. 3.35. (a) TOF mass spectrum of aluminum neutral clusters ionized with an ArF laser. The signal is accumulated over 500 shots, (b) aluminum cluster ions as produced by the LVCS. (From [3.24])

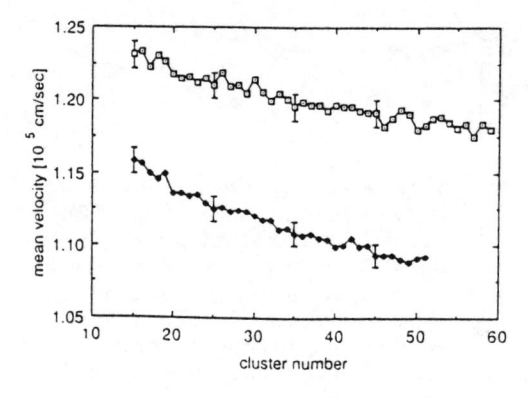

Fig. 3.36. Mean cluster velocities as a function of mass. The curves correspond to two different time delays between the vaporization laser pulse and the trigger of the detector. (From [3.24])

used to burn out a spirographic path on the target [3.106, 3.107]. The disk configuration requires the solution of several problem such as the sealing of the disk-source block interface while keeping the disk moving. Smalley and co-workers [3.108] have used a disk LVCS for producing carbon clusters and C_{60}. A modified version of this source, with a computer-controlled disk motion and a fast pulsed valve has been coupled with a Fourier transform ion cyclotron resonance apparatus [3.109].

3.3.3 Arc Pulsed Sources

Aerosol reactors based on arc discharge can be considered as the archetypes of arc plasma cluster sources. Bulk quantities of small particles can be produced with this apparatus which has become very popular for the synthesis of fullerenes and related materials [3.110, 3.111]. A typical aerosol reactor is shown in Fig. 3.37.

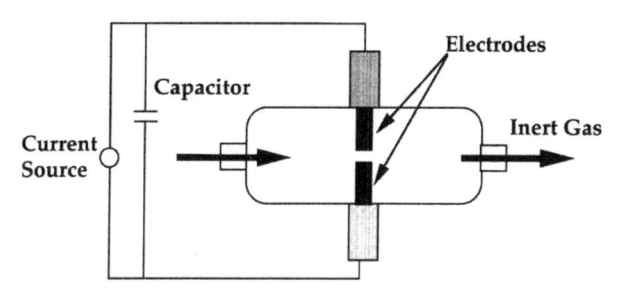

Fig. 3.37. Schematic of an aerosol generator. (From [3.110])

Ni nanoparticles have been produced by striking an arc between a tungsten cathode and a nickel anode and directing a helium gas jet through the arc [3.112]. Particles carried by the gas jet are captured in a liquid nitrogen cold trap. It was found that the helium gas velocity is the predominant factor influencing the size of Ni particles. In particular an increase of helium flow velocity (from $20\,\mathrm{m\,s^{-1}}$ to $56\,\mathrm{m\,s^{-1}}$) results in a reduction of particle size (from 13 nm to 7 nm), due to a reduction of precursor concentration and a higher cooling rate [3.113]. The size distribution of the particles is close to the log-normal distribution.

A similar method is used to produce silicon clusters by spark ablation [3.114]. Material is vaporized from crystalline silicon electrodes using a high energy electric spark. An Ar flux is directed through the electrodes and the clusters nucleate and grow in the flowing argon. Electron microscopy performed on clusters collected on a cold substrate shows the presence of particles with diameters in the 2–4 nm range. Ultrafine gallium nitride powder (20–200 nm diameter) was also synthesized by a dc arc plasma in a gas mixture of nitrogen and ammonia [3.115].

The combination of pulsed valves and arc discharges is a natural evolution towards the production of supersonic cluster beams, allowing a better control of cluster aggregation and producing a highly directional jet. The application of arc discharge to the production of supersonic cluster beams was originally proposed by Meiwes-Broer and co-workers in 1990 [3.116, 3.117]. The pulsed arc cluster ion source (PACIS) was conceived as an alternative to laser vaporization sources: the vaporization of the material to clusterize being obtained by a pulsed high current discharge between two electrodes. The plasma produced in this way is mixed and thermalized by an inert gas pulse generated

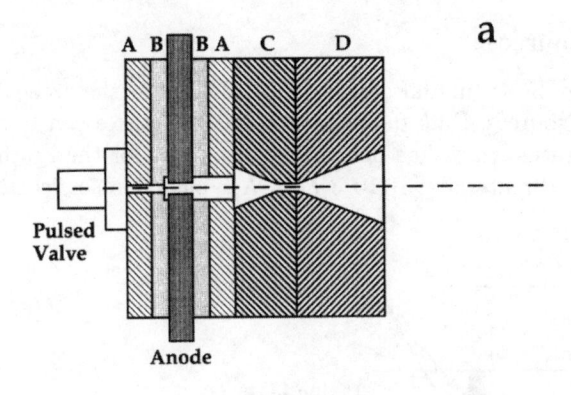

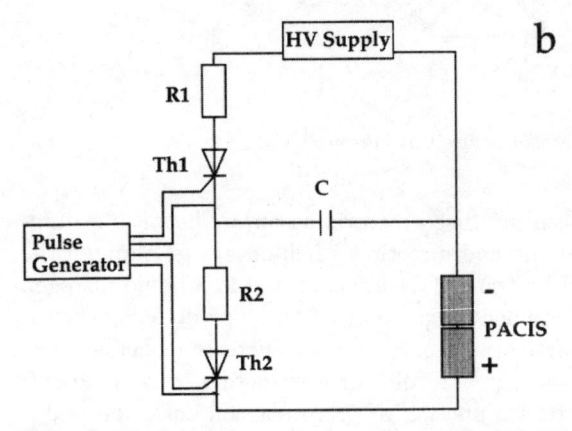

Fig. 3.38. (a) Schematic drawing of the PACIS source: A–aluminum plates, B–insulator block, C–mixing chamber, D–extender with conical nozzle. (b) Power supply using thyristors to initiate the discharge. The capacitor C_1 is charged and discharged by delayed triggering of two thyristors. (From [3.117])

between the electrodes by a pulsed valve. A schematic representation of the PACIS is shown in Fig. 3.38.

Two cylindrical electrodes of a few millimeter diameter are placed in an insulator block (usually Macor or boron nitride), the spacing between the two electrodes is kept at 1–2 mm depending on the material. A pulsed valve supplies pure or mixed gases and its opening can be synchronized with the firing of the discharge. The gas reaches the electrode gap through a channel of variable diameter and length (in the design of [3.117] the channel has a diameter of 1 mm and length of 10 mm). The plasma produced by the discharge and mixed with the carrier gas travels through a larger channel (2 mm diameter, 10 mm length) and it expands into vacuum or it reaches a subsequent "mixing chamber" characterized by a variable volume and shape [3.118, 3.119], which acts as a thermalization zone in a way similar to the one discussed for LVCS.

The discharge driving circuit can be derived from the electronics of an excimer laser where a grounded grid thyratron serves as a high energy switch and yields a maximum of 3×10^{-3} C shot^{-1} at 100 Hz. The applied discharge voltage varying from 2 to 12 kV [3.117, 3.119]. With the PACIS configuration reported in Fig. 3.38, at 5 cm from the nozzle, rates of 0.1 Å/shot and 1.8 Å/shot are obtained for Ag, and Pb respectively, with shot to shot variation of 100% [3.117]. Negative ions up to 2×10^{12} are measured on the skimmer and about 2×10^{11} beyond the skimmer. Positive ions are a factor of three lower compared to negative ions. For Pb clusters a deposition rate of about 5 ML/s (1 monolayer (ML) $\simeq 7 \times 10^{14}$ atoms/cm^2) are achieved at a distance of 10 cm from the source [3.118]. No detailed analysis of the size distribution of the clusters in the beam has been carried out for this source configuration.

Several parameters influence the PACIS performance in terms of produced species, stability, intensity: nucleation and neutralization processes are controlled by the gas dynamics in the source body and by the nature of the discharge.

The geometry of a pulsed arc source can be considered in a way similar to that of LVCS. In particular, channels and volumes inside the source should be designed to avoid any interference between the plasma and the source walls. For an arc source, a further requirement is the electrical insulation of the electrodes from the rest of the source body. Since the most widely used pulsed valves are solenoid actuated, interference between valve circuit and the electrode discharge should be avoided.

The relevant gas pulse characteristics are duration and intensity: the duration determines the quantity of gas introduced into the source chamber and the pressure inside the source. In principle, by varying the delay between the valve opening and the discharge, it is possible to change the gas dynamics between the electrodes and hence to influence the cluster growth. This is of particular importance for source configurations without a thermalization chamber prior to the expansion. The channel connecting the pulsed valve nozzle with the electrode gap can strongly modify the shape of the gas pulse and strongly reduce its peak intensity. To reduce the channel–gas interaction one should use channels as short and as large as possible. In order to prevent plasma interaction with the source walls, the electrode gap region should be carefully designed to find a compromise between plasma expansion and carrier gas pressure.

The thermalization chamber has the same function as in laser vaporization sources, depending on its volume and on nozzle diameter, the residence time of clusters can be varied. Due to the relatively low gas pressure attainable in the thermalization chamber, the expansion does not favor an efficient cooling, however, since the clusters stay several hundreds of microseconds in the source, they can be considered in thermal equilibrium with the source body. This allows time to cool the clusters to very low vibrational temperatures by using cooled sources. In the case of a source with the nozzle placed just after

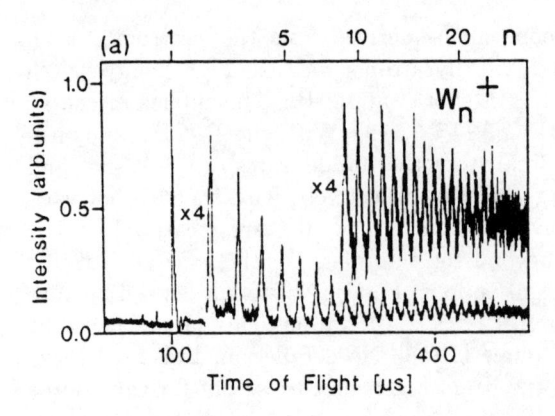

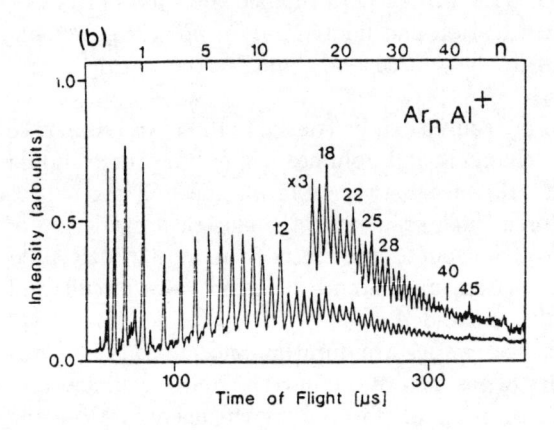

Fig. 3.39. (a) TOF mass spectrum of tungsten cluster ions from a PACIS, (b) aluminum doped argon cluster ions obtained by seeding the aluminum plasma with argon. (From [3.116])

the electrode gap, a relatively efficient expansion regime can be reached and cold ion clusters can be obtained [3.120] (Fig. 3.39). A source with thermalization cavity operates in a gas-aggregation regime (see Sect. 3.2.1), so that the cluster formation process can be described analogously to gas-aggregation sources [3.121, 3.122]. Discharge sources, characterized by a short channel and a converging-diverging nozzle, are more similar to a supersonic source, where a substantial cooling and supersaturation is obtained during the expansion [3.116, 3.120].

The price to pay for higher intensity and lower costs of PACIS, compared to LVCS, is in terms of instability and difficulty of operation. Fast deterioration of the electrodes, clogging, turbulence associated with the discharge in the carrier gas are factors that severely limit a widespread use of PACIS. On the other hand several characteristics, and in particular the high cluster intensities available, make this source an interesting candidate for the generation of cluster beams suitable for deposition. A version of the PACIS which overcomes, at least partially, the reported difficulties, has been proposed allowing the use of this source for cluster beam deposition [3.121].

3.3 Pulsed Sources 87

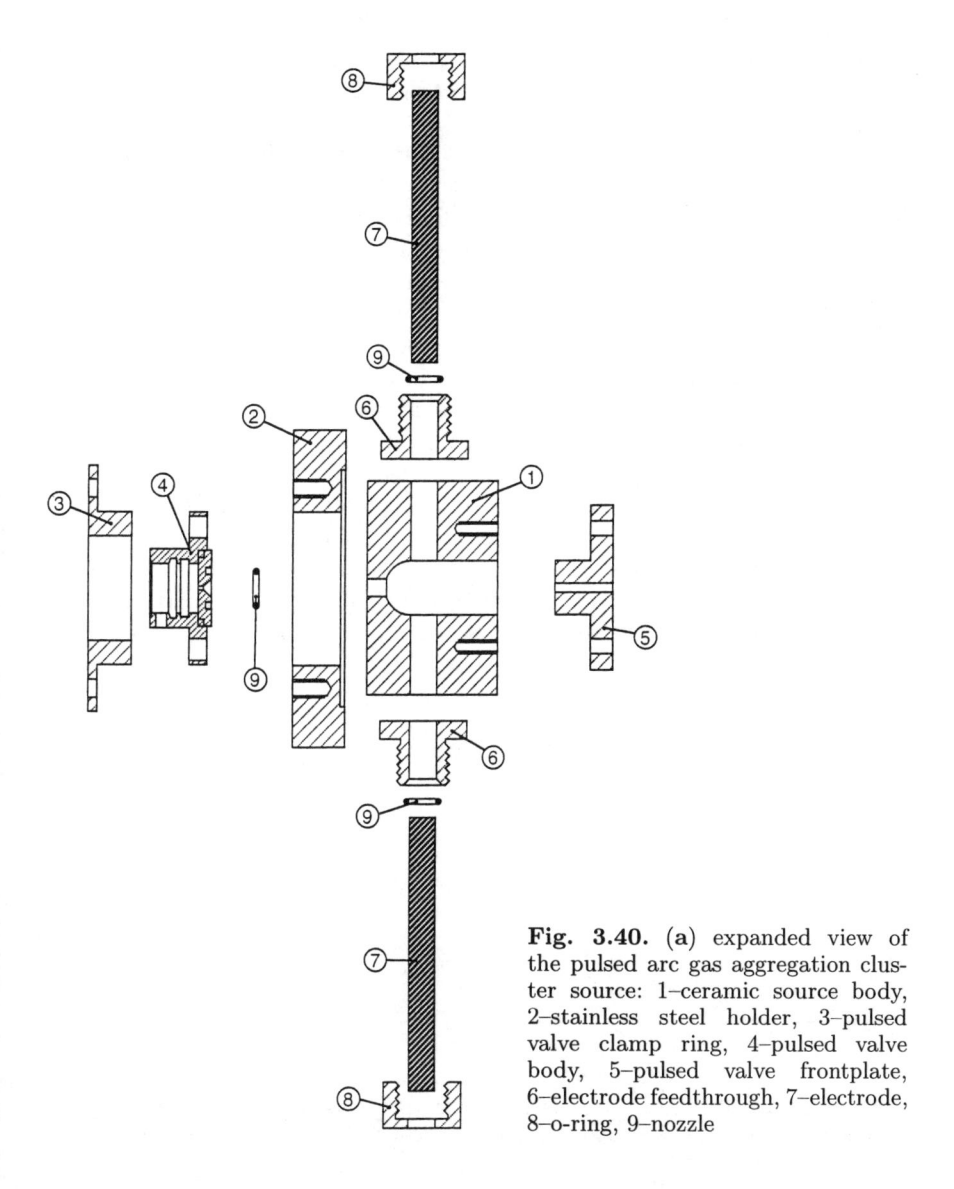

Fig. 3.40. (a) expanded view of the pulsed arc gas aggregation cluster source: 1–ceramic source body, 2–stainless steel holder, 3–pulsed valve clamp ring, 4–pulsed valve body, 5–pulsed valve frontplate, 6–electrode feedthrough, 7–electrode, 8–o-ring, 9–nozzle

A schematic view of the source is shown in Fig. 3.40.

The source is based on a ceramic body (Macor or Shapal-M) mounted on a stainless steel support for the alignment with the skimmer. The pulsed valve is fixed on the rear side of the source body by a clamping ring allowing a careful alignment and seal of the frontplate. This is very critical for prolonged operation of the source and stable performances. A few millimeter long channel connects the valve with the vaporization cavity that has a cylindrical shape with one end rounded. This configuration favors both plasma–gas mixing and cluster condensation, while assuring stable operation. The volume of the cavity can be varied by changing the removable nozzle. Particular care should also be taken to seal the electrodes to prevent their erosion on unwanted positions.

With this source, the mass distribution of clusters ions (cations) is substantially different from that of neutrals: 0–1200 atoms/cluster (with mass distribution peaked at 350 atoms/cluster) for ions, against 0–3000 atoms/cluster (peaked at 750 atoms/cluster) for neutrals. This is probably due to a trade-off between cluster growth and charge neutralization: the growth of large clusters requires long residence times inside the source, but this also implies a higher chance to be neutralized. These results suggest that the frequently made assumption on the correspondence of neutral and ion cluster beams produced by the same source must always be carefully verified.

Depending on the conditions of the expansion, deposition rates from several nm/min up to 60 nm/min can be routinely obtained. A fraction of about 10% of the total average flux is due to anions, while cations account for about 2%. Neutral cluster velocities are in the range of 1400–1800 m s^{-1}, depending on the exit time. The typical kinetic energy of a medium-size cluster is thus about 0.3 keV. The velocity of the cluster ions is spread over the range 1200–1900 m s^{-1}, the distribution peaking at 1700 m s^{-1} (Fig. 3.41).

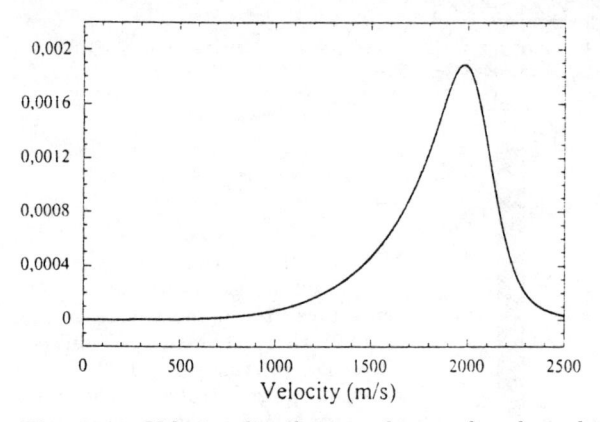

Fig. 3.41. Velocity distribution of neutral carbon clusters. This curve has been obtained by summing the time of flight of clusters with different residence times and hence different velocities. (From [3.122])

The cluster mass distribution can be varied by changing the geometrical parameters of the source. In particular, the residence time and the pressure in the thermalization chamber can affect the growing processes favoring or preventing the formation of large aggregates. In Fig. 3.42 the evolution of the center of mass distribution for three different nozzles is shown. The nozzle diameter influences the gas dynamics in the source and it is a parameter that allows a rough control of the cluster size distribution.

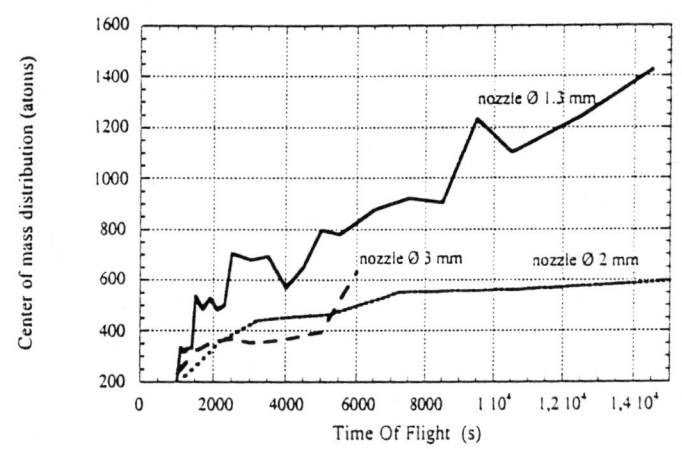

Fig. 3.42. Evolution of the center of mass distribution for three different nozzle diameters. (From [3.122])

4. Characterization and Manipulation of Cluster Beams

4.1 Mass Spectrometry

4.1.1 Quadrupole Mass Spectrometry

A quadrupole mass spectrometer is based on the original instruments developed by Paul and co-workers [4.1–4.3] where the selection of the ions is carried out, according to their mass to charge ratio (m/q), in a quadrupole rf electric field so that only ions of a defined mass will reach a suitable detector. Mechanically it consists of four rod-shaped electrodes with an hyperbolic cross-section. Figure 4.1 shows the end-on view of this type of analyzer. The opposite electrodes are electrically connected and the voltage applied consists of two simultaneous components: one is a constant voltage U and the other one is a radio frequency (rf) $V_0 \cos(\omega t)$. The system of coordinates that we assume and the values of the potentials on the four electrodes is also shown in the figure.

An ion entering the gap between the electrodes, delimited by a circle of radius r_0 ($2r_0$ being the distance between two opposite electrodes), will experience a field that may be written in the form:

$$\phi = \frac{(U + V_0 \cos \omega t)(x^2 - y^2)}{r_0^2} . \tag{4.1}$$

The basic equations of motion for an ion during its flight into the quadrupole can then be written in the form:

$$m \frac{d^2 x}{dt^2} + \frac{2q(U + V_0 \cos \omega t) x}{r_0^2} = 0 , \tag{4.2}$$

$$m \frac{d^2 y}{dt^2} - \frac{2q(U + V_0 \cos \omega t) y}{r_0^2} = 0 , \tag{4.3}$$

$$m \frac{d^2 z}{dt^2} = 0 . \tag{4.4}$$

These equations can be rewritten in a more convenient form by defining three dimensionless parameters:

$$A = \frac{8qU}{mr_0^2 \omega^2} , \tag{4.5}$$

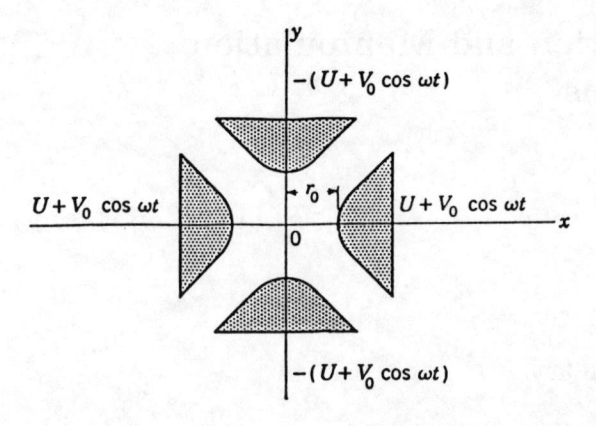

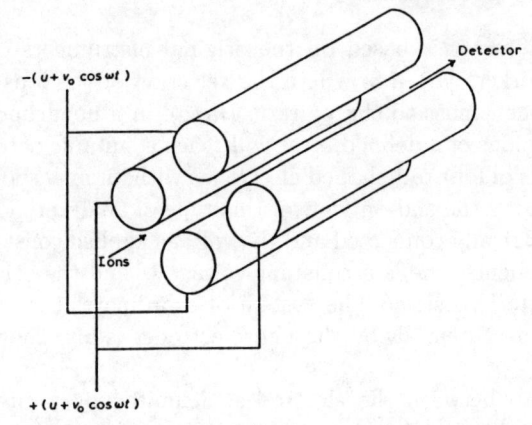

Fig. 4.1. *Top*: end-view of an ideal quadrupole made of 4 electrodes with hyperbolic cross-section. *Bottom*: a real quadrupole made by 4 cylindrical rods. The system of coordinates used is also shown

$$Q = \frac{4qV_0}{mr_0^2\omega^2}, \tag{4.6}$$

$$\varsigma = \frac{\omega t}{2}. \tag{4.7}$$

With this change of variables one obtains the following equations (Mathieu equations):

$$\frac{d^2 x}{d\varsigma^2} + (A + 2Q\cos 2\varsigma)x = 0, \tag{4.8}$$

$$\frac{d^2 y}{d\varsigma^2} - (A + 2Q\cos 2\varsigma)y = 0. \tag{4.9}$$

A detailed discussion of ion trajectories, solutions of the Mathieu equations, is given by Dawson [4.4]. The solutions describe the motion of the ion as oscillating in the x and y direction while traveling in the z direction. Depending on the values assumed by the parameters A and Q, the oscillations may be stable and characterized by an amplitude confined to the gap between the electrodes at any time t. Otherwise they may become instable with amplitude increasing exponentially with the ion impinging on the electrodes or flying away from the gap. The quadrupole works then as a band/pass filter that is tunable by properly changing the parameters U, V_0, and ω.

From the Mathieu equations a stability diagram can be derived to describe the regions of values of A and Q that give rise to stable ion trajectories. It is important to note that these conditions are independent of the initial position and velocity of the ion and are not affected by the phase of the rf potential at the injection time. Figure 4.2 shows such a diagram where a line $A/Q =$ constant is also drawn.

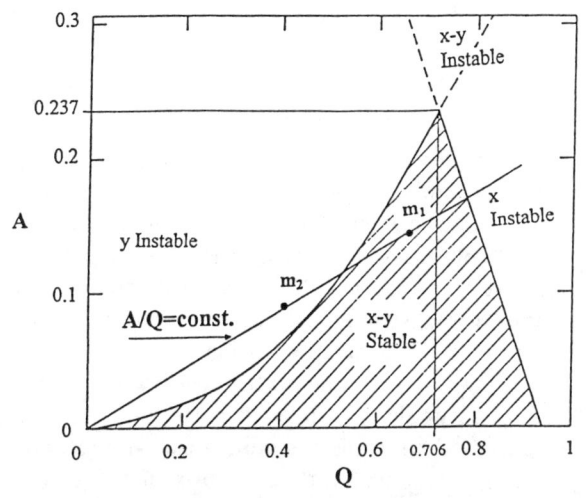

Fig. 4.2. Stability diagram of the solutions of the Mathieu equations. The dashed area corresponds to the values of A and Q that give rise to stable ion trajectories. The ion masses m_1 and m_2, lying in two different regions, will have a stable oscillating trajectory and, respectively, an unstable diverging trajectory. The result is that only the ion m_1 will pass through the quadrupole

The slope of this line is given by the ratio U/V_0. Its points are representative of the ion masses and it can cross the region of stable oscillations depending on the value of U/V_0. In Fig. 4.2 two different ions of mass m_1 and m_2 are reported on the A/Q constant line. One lies inside the region of stable oscillations and the other outside it. For masses like m_1, the ions follow an oscillatory path while traveling through the quadrupole whereas for other masses, such as m_2, the ions will increase their distance from the center of the quadrupole and will be eventually lost because of their lateral deflection. Because of the characteristic shape of the stability region in the $A-Q$ plane, the line, depending on the specific value of the constant A/Q, could span this region crossing it at two points that define the lower and upper mass of the transmitted range. This will define the theoretical resolution of the quadrupole that would reach the maximum (theoretically infinite) when the line A/Q crosses at the apex ($A = 0.237, Q = 0.706$) the cuspid that characterizes the stability region. In this limit condition the transmitted range reduces to a single mass (at least in theory).

A mass spectrum is obtained maintaining the fixed ratio A/Q and by varying ω_0 or, more commonly, the values of U and V_0. A typical arrangement is to chose a linear correlation, $U = aV_0 + b$. In this case the performance of a quadrupole may be described by the following simple formulas [4.5] that calculate the peak m and the width Δm of the mass band in atomic mass units:

$$m = \frac{1.385 \times 10^7 V_0}{(\nu r_0)^2}, \tag{4.10}$$

$$\frac{\Delta m}{m} = 7.936 \left((0.16784 - a) + \frac{b}{a} \right), \tag{4.11}$$

where $\nu = 2\pi\omega$ is the rf frequency. The relative choice of a and b define the theoretical mode of operation of the quadrupole.

In real instruments the performance is strongly affected by the design: diameter of input and output of collimators, profile of the rods (hyperbolic profiles, very difficult to make, are often replaced by cylindrical ones), length of the quadrupole, precision and tolerances in machining and mounting, quality of electronics, etc. The combined effect of all such factors on resolution can be easily checked by measuring intensity and width of a few peaks as a function of the ratio U/V_0. A quadrupole with a good performance should show a decrease of both intensity and width as the value of U/V_0 increases. Otherwise, a depletion in intensity combined with no improvement in resolution is a clear indication that the mechanical and electronic imperfections strongly affect the instrument. The mechanical construction may strongly limit the performance of the spectrometer. For example, one major factor limiting resolution is the rod manufacturing accuracy, whose effects can be calculated by the formula:

$$\left(\frac{\Delta m}{m} \right)_{mech} = \frac{2\Delta r_0}{r_0}, \tag{4.12}$$

where the key parameter limiting resolution is the tolerance Δr_0 on the radius over the whole length of the quadrupole electrodes. This consideration favors the choice of large rod diameters but the need of strongly increasing rf power (proportional to $\nu V_0^2 = m^2 \nu^5 r_0^4$) places a serious constraint in this direction. Commercially available systems have typical values of $(\Delta m/m)_{mech} = 2 \times 10^{-4}$.

The limitations arising from the electronics are mainly caused by frequency and amplitude instabilities that make the choice of good quality electronics critical to the final performance of the mass spectrometer. The quality of the environment where the quadrupole is used is also important since the growth of insulating layers on the electrodes may give rise to field changes beyond the control of the electronics. The only solution is to use care in cleaning and handling the device. If used in UHV a good bakeout can often solve this kind of problem.

The finite size of the input and output collimators of the spectrometer further depletes the performance of the spectrometer. The fact that the ions can enter the quadrupole field with different initial energies, combined with the finite length of the electrodes, reduces the filtering capacity of the instrument. The initial position in the (xy) plane is delimited by the area of the collimator and allows ions with transversal energies to enter the quadrupole. In fact, it is quite important to prepare the ion beam to be filtered by the quadrupole with characteristics as close as possible to the ideal in terms of energy and angle of entrance. Dawson [4.4] gives the following expressions correlating longitudinal E_z and transversal E_{xy} energies, diameter of the input collimator φ, and length of the quadrupole L:

$$D < r_0 \sqrt{\frac{\Delta m}{m}}, \tag{4.13}$$

$$E_{xy} < 5 \times 10^{-9} m \nu^2 r_0^2, \tag{4.14}$$

$$E_z < 4 \times 10^{-10} m L^2 \nu^2, \tag{4.15}$$

where L is in meters, mass in amu, the frequency in Hz and the energy in eV. Computer simulations [4.6, 4.7] confirm how important is the angle of entrance of the ions and the rf phase in the transmission probability of real quadrupoles of finite length.

The effects of fringe fields, present in the regions where both the continuous and the rf fields change from zero to the values set into the quadrupole, are also important and affect, in particular, ions with low energies. In terms of the stability diagram this corresponds to the fact that the ions have to cross a region of instability (as shown in Fig. 4.2), where the two fields change could disperse them. The net result is a more or less severe cut in transmission in the high mass range [4.8] of the quadrupole. One way to partially overcome this problem is to spatially decouple the onset of the dc field from that of the rf. This can be done by introducing a short quadrupole, immediately before a standard quadrupole, to which only the rf is applied. In this way, as can be easily seen in the stability diagram of Fig. 4.2, since A is zero in this configuration, the ions entering the filter will not cross a region of instability.

Other critical aspects that can easily reduce the transmission and the resolution of a quadrupole derive from electrical tuning and balance of the quadrupole. The major problems usually arise from differences in the fields applied to the two opposite electrodes that in an ideal quadrupole are assumed equal. Commercially available instruments are usually provided with electronic adjustments devoted to compensate these effects. In more recent instruments, the use of computer-controlled electronics allows for an automatic control of the different adjustments needed and for the fine control of the voltages U and V_0 in the mass scans. Quadrupoles with unit mass resolution up to masses larger then 5000 amu and mass selection up to more then 8500 amu are reported [4.9].

One of the advantages of quadrupoles as mass spectrometers is that they are compact compared to other devices. On the other hand, they have a limited range of mass if their size and rf power requirements are kept in a reasonable range.

Mass quadrupoles are inherently continuous devices with high transparency so that their ideal use is when coupled to continuous beams. They may represent the best choice both for detection and mass selection of relatively small clusters. An example are the experiments of Vajda et al [4.10] on small Ni aggregates. They selected clusters in size (up to 31 monomers) using a quadrupole mass spectrometer and then let them react with CO, analyzing the products again with a quadrupole filter. Another interesting example are the experiments on bimetallic clusters [4.11], such as Nb_mAl_n, V_mAl_n with n up to 31, formed by a dual laser vaporization source, that are characterized by a quadrupole mass spectrometer. Another common field of application is the characterization of molecular beams of gas clusters [4.12].

The use of quadrupoles as filters and ion guides for size-selected cluster beam deposition is briefly discussed in Sect. 4.3.2.

4.1.2 Time-of-Flight Mass Spectrometry

Time-of-flight mass spectrometry (TOF/MS) is the technique of choice for studying clusters in molecular beams [4.13]. The operating principle of a TOF spectrometer is very simple: ions with the same kinetic energy but with different masses must have different velocities (Fig. 4.3). When injected into a drift tube the ions will separate according to their masses. TOF/MS has several advantages over other mass spectrometric techniques: it is inexpensive and easy to build, the ion transmission is almost unitary over an almost unlimited mass range, a complete mass spectrum can be recorded every few microseconds and it can be efficiently coupled to pulsed sources [4.14].

The rapid growth of cluster physics and the development of pulsed cluster sources have stimulated many technical refinements to improve the TOF performance in terms of mass resolution and sensitivity [4.15–4.17].

The mass resolution of a TOF/MS can be defined as the largest mass, M, for which adjacent masses are essentially completely separated, or, in other words, the mass for which the time spread of isobaric ions just equals the time separation between two adjacent mass peaks [4.14].

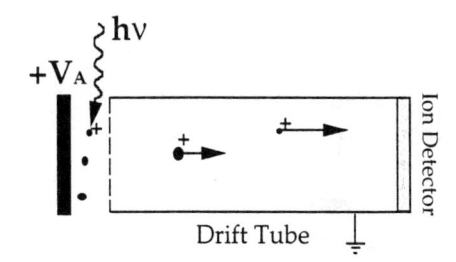

Fig. 4.3. TOF/MS operating principle

If all the ions are formed in a plane parallel to the accelerating electrodes and with no initial velocities, the ions flight time will depend only on the charge to mass ratio. The resolution would be limited only by factors such as inhomogeneities in the electric fields, temporal spread in the ionization source, and response speed of the detection electronics. However, under normal experimental conditions, the variation of ions TOF is dominated by the spatial distributions and initial ion velocities. The spatial spread is due to the finite dimensions of the ionization source (laser or electron spot): isobaric ions starting at different points will reach the detector at different times. The time spread, introduced by initial velocities, is due to the fact that isobaric ions produced at the same position but with different velocity components along the spectrometer axis, will arrive at different times since the velocity components must be compensated. The velocity spread is related to the initial thermal energy of the particles having a Boltzmann distribution.

The resolving power of the TOF/MS depends on the ability of reducing the effects of the initial spatial and energy spreads of the ions. Wiley and McLaren [4.14] have shown that, for a given configuration of the accelerating fields, the effects of the initial spatial width of the ion packet can be minimized by fulfilling the condition:

$$\frac{dT}{ds} = 0, \tag{4.16}$$

where T is the time of flight and s the initial position of the ions [4.14].

They proposed a class of TOF spectrometers, widely used in cluster physics, consisting of two accelerating regions separated by a grid, a drift region and an ion detector (Fig. 4.4). Ions formed or collected in the first region are accelerated through the drift tube and then detected.

The double field configuration allows more flexibility introducing, as parameters, the dimension of the second acceleration zone d and the electric field ratio E_d/E_s. In this way it opens the possibility to partially correct for the initial space and energy distributions. Space resolution can be improved by reducing the initial spatial spread or by giving to each ion with the same mass a different velocity depending on its initial position. The energy spread can be reduced by increasing the ratio of the ion total energy to its initial energy. Energy focusing can also be obtained by introducing a time lag between ion creation and ion acceleration. During this time lag, some of the

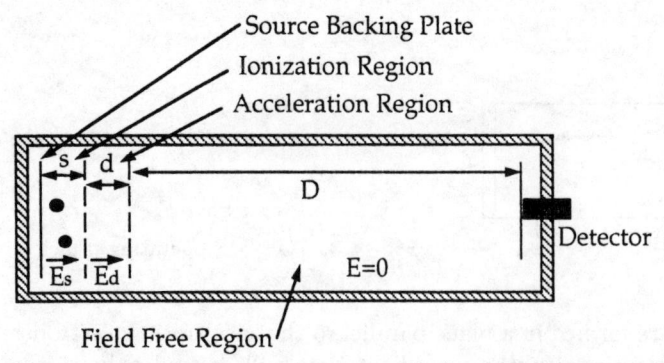

Fig. 4.4. Basic geometry of a linear TOF/MS as proposed by Wiley and McLaren. (From [4.14])

ions move towards new positions which have the proper flight times to correct for the initial velocity [4.14]. Space focusing and energy focusing require opposite conditions for several parameters of the spectrometer, therefore the best resolution is a compromise between space and energy resolution. It has been proven analytically that the Wiley–McLaren configuration achieves the highest possible energy focusing. It is thus the best choice if space and energy focus are required at the same time [4.18].

Despite its many advantages the two-stage linear TOF/MS, has a resolution m/δm which rarely exceeds a few hundred. A considerable improvement in resolution can be achieved by adding a stage of deceleration-acceleration of the ion bunch in order to correct for the energy spread [4.19], this configuration, called reflectron TOF/MS, is discussed below.

Another method to improve the resolution is to narrow or even eliminate the velocity spread of the molecules: this can be achieved by preparing the species under study in a supersonic molecular beam travelling perpendicular to the spectrometer axis (Fig. 4.5) [4.20, 4.21]. As already discussed in Chap. 1 the kinetic energy perpendicular to the beam axis is very low. TOF/MS resolution can thus be considerably increased by improving only space focusing.

Several configurations of linear TOF/MS have been proposed in order to achieve high resolution when coupled to molecular beam systems producing clusters [4.13, 4.15, 4.17, 4.22–4.24]. In particular in [4.22, 4.24] a two-stage mass spectrometer characterized by a high resolution with a second order space focus has been proposed. The fulfillment of the condition:

$$\frac{d^2 T}{ds^2} = 0 \tag{4.17}$$

allows a good compensation of the initial spatial distribution over a large volume.

Since the designs of a TOF/MS are inherently scalable [4.25], a parameterization allows us to write analytical expressions to calculate the TOF

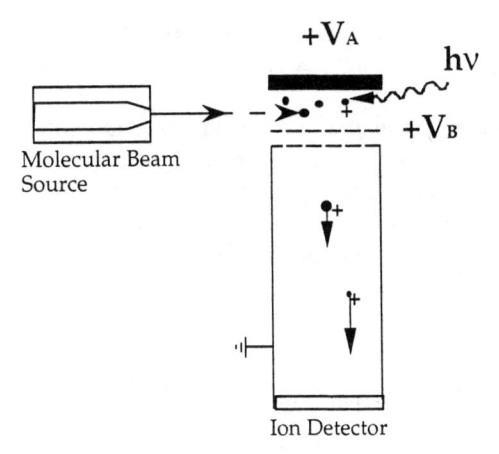

Fig. 4.5. Schematic representation of the coupling of a molecular beam with a TOF/MS

parameters as a function of one single variable, namely the ratio of the voltages of the two accelerating stages. The parametrization is well suited to analyze the performance of the spectrometer and of the factors limiting the resolution.

Scheme of a Two-Stage TOF Mass Spectrometer. Figure 4.6 shows a schematic representation of a typical two-stage TOF mass spectrometer. It can be divided into four regions:

(i) The ionization region, where particles are ionized and from which they are extracted by the electric field E_1 (pulsed or static).

(ii) The second region is an accelerating zone where the ions experience a static field E_2 which brings their kinetic energy to a final value U_0. The role played by this second field is to introduce a sufficient number of parameters which can be adjusted in order to produce a second order space-focusing.

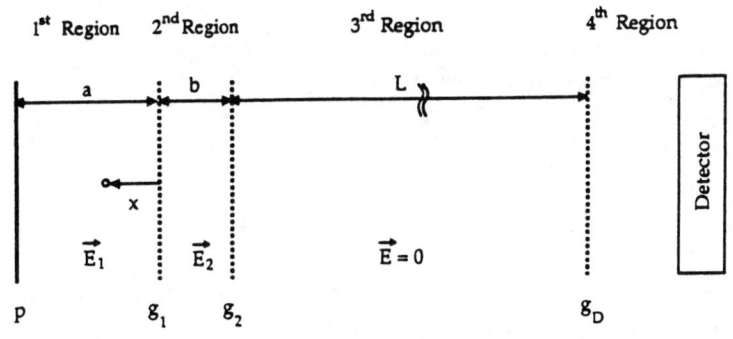

Fig. 4.6. Scheme of a two-stage TOF/MS. The spectrometer is composed of four regions separated by grids. Lengths of the first three zones are indicated by a, b and L in the figure as well as in the text. (From [4.24])

(iii) The third region is a field-free zone where different ions separate since their speed is proportional to the inverse square root of the mass to charge ratio m/z.

(iiii) The fourth region spans between the end of the field free region and the detector. Since the time spent by the ions in this zone is generally very short compared to the total time of flight, its influence can be compensated by a small adjustment of the accelerating potentials.

The time of flight of an ion of charge z and mass m is the sum of the times spent in regions 1, 2 and 3; that is

$$T(x, v) = \frac{m}{zE_1} \left(\sqrt{v^2 + \frac{2z}{m} V_e \frac{x}{a}} - v \right)$$

$$+ \frac{m}{zE_2} \left(\sqrt{v^2 + \frac{2z}{m} \left(V_e \frac{x}{a} + V \right)} - \sqrt{v^2 + \frac{2z}{m} V_e \frac{x}{a}} \right)$$

$$+ \frac{L}{\sqrt{v^2 + \frac{2z}{m} \left(V_e \frac{x}{a} + V \right)}}, \tag{4.18}$$

where x is the distance of the ion initial position from grid g_1 (we use the same sign convention of [4.14]), v is the ion initial velocity which is negative when directed towards the grid g_1, V_e and V are the potentials generating the extraction field E_1 and the accelerating field E_2 respectively (see Fig. 4.6 for the meaning of other symbols used in this section).

Introducing the dimensionless parameters $\beta = V/V_e$, $\delta = b/a$, $\lambda = L/a$ and $s = x/a$, and dropping the dependence from v (that is $v = 0$), equation (4.18) can be rewritten as:

$$T(s) = K \left[\left(1 - \frac{\delta}{\beta}\right) \sqrt{s} + \frac{\delta}{\beta} \sqrt{s + \beta} + \frac{\lambda}{2\sqrt{s + \beta}} \right], \tag{4.19}$$

where K is

$$K = a \sqrt{\frac{2m}{zV_e}}, \tag{4.20}$$

and s covers the whole ionization region spanning the interval [0,1].

The first and the second derivative are:

$$\frac{dT(s)}{ds} = \frac{K}{2} \left[\left(1 - \frac{\delta}{\beta}\right) \frac{1}{\sqrt{s}} + \frac{\delta}{\beta} \frac{1}{\sqrt{s + \beta}} + \frac{\lambda}{2} (s + \beta)^{-\frac{3}{2}} \right], \tag{4.21}$$

$$\frac{d^2T(s)}{ds^2} = \frac{K}{4} \left[\left(1 - \frac{\delta}{\beta}\right) s^{\frac{3}{2}} + \frac{\delta}{\beta} (s + \beta)^{-3/2} + \frac{\lambda}{2} (s + \beta)^{-\frac{3}{2}} \right]. \tag{4.22}$$

First and second order space focusing is obtained by choosing the parameters β, δ, and λ so that (4.16) and (4.17) are verified.

In order to obtain simple analytical expressions we assume that $s_0 = 1/2$, i.e. the midpoint of the ionization region is the starting point for the ions to be mass selected. We note that this is by no means a limitation on the possible geometry, because in any Wiley–McLaren TOF mass spectrometer, the first stage can be reshaped without modifying the focusing conditions, provided that electric field E_1 and distance x_0 are not affected by the changes. Space focusing is thus achieved when:

$$\left(1 - \frac{\delta}{\beta}\right) + \frac{\delta}{\beta}(1 + 2\beta)^{-1/2} - \lambda(1 + 2\beta)^{-3/2} = 0, \qquad (4.23)$$

$$\left(1 - \frac{\delta}{\beta}\right) + \frac{\delta}{\beta}(1 + 2\beta)^{-3/2} - 3\lambda(1 + 2\beta)^{-5/2} = 0. \qquad (4.24)$$

These equations can be solved keeping β as a free parameter, giving:

$$\delta = \frac{\beta(\beta - 1)\sqrt{1 + 2\beta}}{1 + \beta(\beta - 1)\sqrt{1 + 2\beta}}, \qquad (4.25)$$

and

$$\lambda = \frac{\beta(1 + 2\beta)^{3/2}}{1 + (\beta - 1)\sqrt{1 + 2\beta}}. \qquad (4.26)$$

Once the dimensions of the first accelerating zone are fixed, (4.25) and (4.26) give the lengths of the second accelerating region and of the free flight zone, respectively, expressed in relative units. A solution is obtainable for any given value of parameter β in the domain $(1, +\infty)$.

Although the initial energy spread can be significantly reduced by the coupling with a molecular beam, it is interesting to consider the effect of the residual energy spread on the choice of the configuration and of the construction parameters.

The initial energy distribution has a considerable effect in the first stages of acceleration since the ions have not yet reached their final energies. For this reason, one must minimize the fraction of total time of flight that ions spend in the acceleration zone. The following expression gives this fraction of time as a function of β

$$\frac{2\beta - 1}{4\beta}. \qquad (4.27)$$

Figure 4.7 shows the branch of hyperbola given by (4.27). Choosing β close enough to 1, the time spent in the first two zones approaches 25% of the total time of flight.

In most cases, for ions with the same m/z ratio, the contribution of the initial energy to the spread of the arrival time is dominated by the turn-around time [4.14]. This is the case of two identical ions starting in the same position, with the same initial speed but with opposite directions (namely towards and away from the detector). The time $\Delta T_{\text{t.a.}}$ needed to decelerate the counter propagating ion and to accelerate it toward the detector represents the difference between the times of flight of the two ions. It is given by

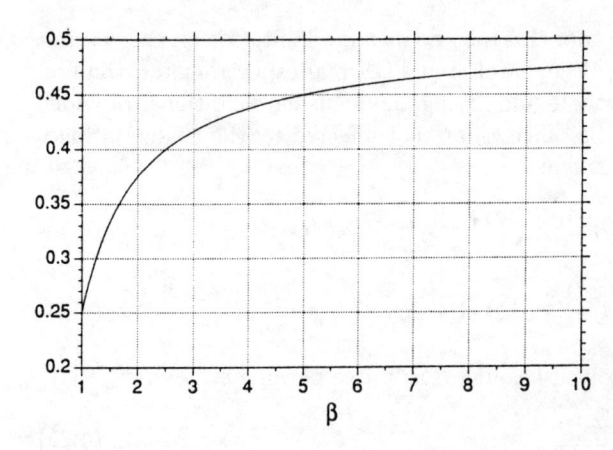

Fig. 4.7. Ratio between the time spent by an ion in the first two regions and the total time of flight as a function of the parameter β. (From [4.24])

$$\Delta T_{\text{t.a.}} = \frac{2amv_0}{zV_e} \tag{4.28}$$

where v_0 is the initial velocity. In order to evaluate the effect of the turn-around time on the final resolution, an important parameter is the ratio between total time of flight and turn-around time, for a given value of the final kinetic energy U_0. We have:

$$\frac{T}{\Delta T_{\text{t.a.}}} = \sqrt{\frac{U_0}{2m}} \frac{1}{2v_0} \left(\frac{4\beta^2}{\sqrt{1+2\beta}\left(1+(\beta-1)\sqrt{1+2\beta}\right)} \right). \tag{4.29}$$

A plot of (4.29) is shown in Fig. 4.8. Large values of the ratio are obtained for small β, although $T/\Delta T_{\text{t.a.}}$ remains within about 90% of its maximum value up to $\beta = 7$.

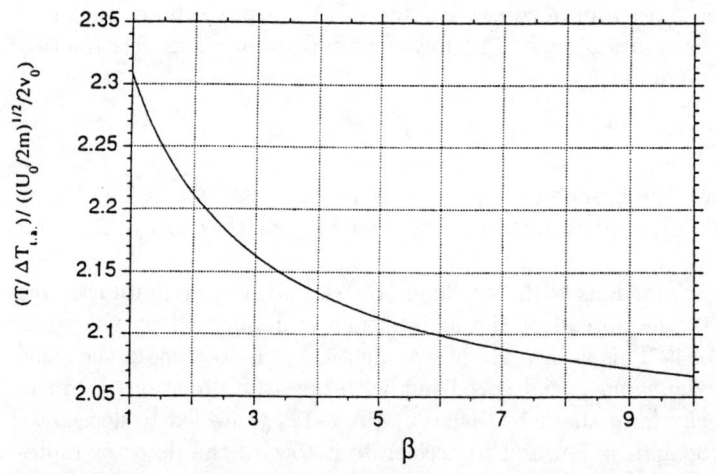

Fig. 4.8. Plot of (4.29). The term in braces is plotted versus β. (From [4.24])

For $\beta = 1.133$ we have $\delta = 0.22$ and $\lambda = 5.4$; if the first accelerating zone has a length $a = 5$ cm, the second one is 1.1 cm long and the free flight zone is 27 cm. In Fig. 4.9 a plot of (4.19) with these values of the parameters is shown. The standard deviation for the times of flight of identical ions with starting position spread over a 1 cm wide region has been computed: the ratio between the mean time of flight and standard deviation is $\langle T \rangle / \sigma_{\langle T \rangle} = 2.4 \times 10^4$.

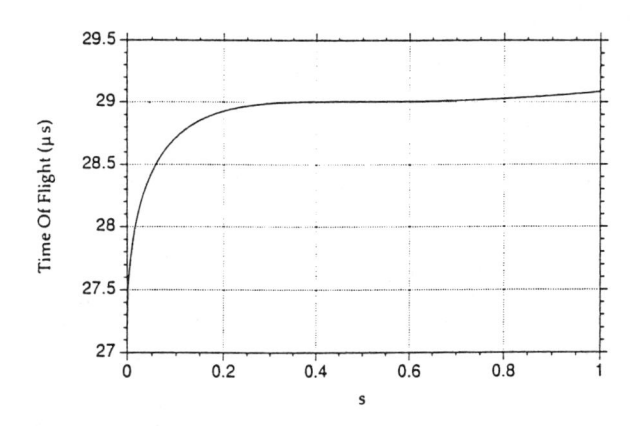

Fig. 4.9. Time of flight of an ion of mass $m = 1000$ amu as a function of the starting position inside the first region. The simulated instrument has a β value of 1.133 with a total length of 33.1 cm. The kinetic energy U_0 given to the ion by the electric field is 1 keV. (From [4.24])

This does not represent the obtainable resolution because in real instruments other factors affecting resolution must be taken into account. Nevertheless $\langle T \rangle / \sigma_{\langle T \rangle}$ is a measure of the degree of space focusing that can be achieved and it is related to the maximum width of the ionization volume which can be used without losing resolution. In Fig. 4.10, we report an analysis of $\langle T \rangle / \sigma_{\langle T \rangle}$ as a function of β, showing that a small value of β is the best choice.

It should be noted that other effects can limit the overall resolution of a mass spectrometer. The duration of the ionizing pulse is, for example, a limiting factor. Considering two cluster ions produced at the beginning and at the end of a laser ionization pulse τ_i (usually of several nanoseconds), the difference in flight time will be equal to τ_i [4.21, 4.22]. This effect can be corrected if a pulsed acceleration voltage is used to extract the ions from the first region of the spectrometer, the limiting factor then being the rise time of the extraction potential.

In general all the limiting effects will sum to give the total time width:

$$\delta T = \sqrt{\sum_i (\delta T_i)^2}, \tag{4.30}$$

affecting the overall resolution and giving:

$$\left(\frac{m}{\delta m}\right) = \left(\frac{T}{2\delta T}\right) = \left[\sum_i \left(\frac{\delta m}{m}\right)_i^2\right]^{-1/2} = \left[\sum_i 2\left(\frac{\delta T}{T}\right)_i^2\right]^{-1/2}. \tag{4.31}$$

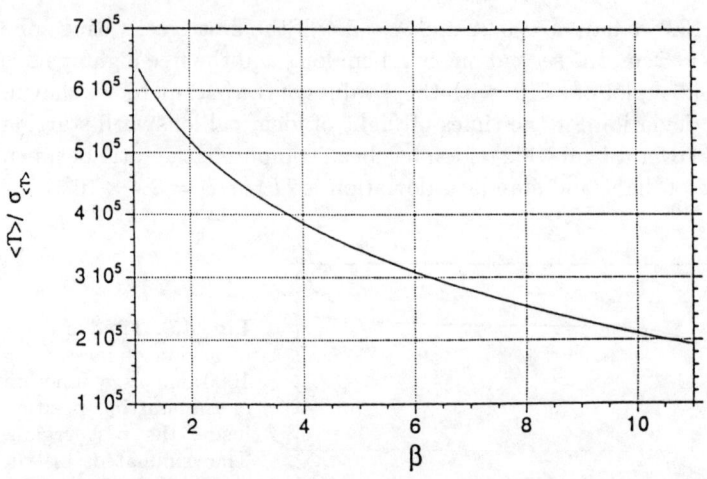

Fig. 4.10. Ratio between the mean time of flight and its standard deviation as a function of β. The overall length of the resulting geometry is kept fixed at 1 m and total kinetic energy U_0 to 1 keV. Ions are assumed to be uniformly spread over a 1 cm wide zone. (From [4.24])

The global mass resolution is usually determined by the effect predominant on the mass peak widths [4.22].

Monte Carlo simulations show that a mass resolution up to 1000 should be obtained with a relatively short (~30 cm) linear instrument [4.24]. Figure 4.11 represents the simulated peak shapes of two ions of unitary charge with masses equal to 1000 and 1001 atomic masses respectively. Ions are randomly generated with starting positions spread over a 1 cm wide normal distribution and with initial kinetic energies distributed according to a temperature of 10 K. The separation between the peaks is almost the same as the peak width, which is about 8 ns; if the ion formation time is shorter (as it could be by ionizing with short laser pulses), and fast detection electronics are available, a resolution close to 1000 can be obtained.

This is a highly idealized situation difficult to obtain for cluster beams, where the initial energy spread is a serious problem only partially eliminated by the use of supersonic expansions. Since the turn-around time δT_v for a velocity v is given by $\delta T_v = T(x-v) - T(x+v)$, it can be shown that:

$$\delta T_v = \frac{2mv}{qE}. \qquad (4.32)$$

It is then evident that small particles in a perpendicular molecular beam can have a very small turn-around time since their velocity dispersion is strongly reduced on the spectrometer axis. However for large particles the turn-around time can seriously affect the resolution. The use of very high extracting fields can reduce the turn-around time.

High extracting fields can also be beneficial for the correction of the initial velocity components of the particles when the spectrometer is used perpen-

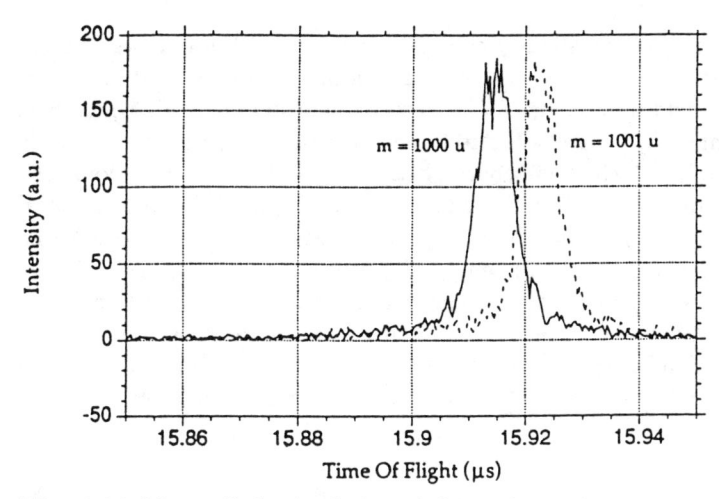

Fig. 4.11. Monte Carlo simulation of the peaks resulting from ions of mass 1000 and 1001 spread over a 1 cm (FWHM) wide zone with normal distribution. Initial velocity has been also generated according to a beam temperature of 10 K. The simulated spectrometer has a total length of 33.1 cm. The total kinetic energy given to the ions is 3 keV. (From [4.24])

dicular to the beam. Since the clusters have translational velocities close to that of the carrier gas, their initial energy can be quite large thus strongly reducing the size range which can be steered onto the detector. This problem can be solved by using parallel deflection plates ovelaying the entire field-free flight zone. A simple voltage step applied to the plates with a properly adjusted time delay performs very well in steering ions over a wide mass range onto the detector, without significantly affecting the final resolution [4.17].

Reflectron TOF/MS. The reflectron mass spectrometer (RTOF/MS) has been proposed to correct the energy spread of the ions. A schematic representation of a RTOF/MS, as originally presented in [4.19, 4.26] is shown in Fig. 4.12.

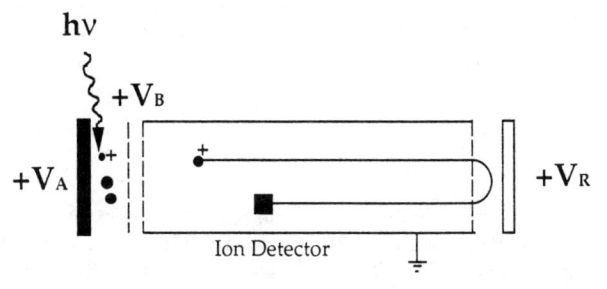

Fig. 4.12. Schematic representation of a RTOF/MS

The configuration can slightly change depending on the number of focal points in the spectrometer. An ion bunch produced in the ionization region travels along a field-free region before entering a system of grids where it is slowed down and then reflected. After reflection it is reaccelerated towards the detector from which it is separated by a second field-free region. The variation of transit time through the two field-free regions, due to the initial energy spread, is compensated by the different residence times of ions in the decelerating-reflecting gaps. More energetic ions penetrate more deeply in the grid region thus spending more time compared to the less energetic ions. This scheme realizes a second order energy focusing. Using a reflectron in series with a two-stage linear spectrometer, one can obtain a focusing of the spatial and energy spreads and a considerable increase of mass resolution.

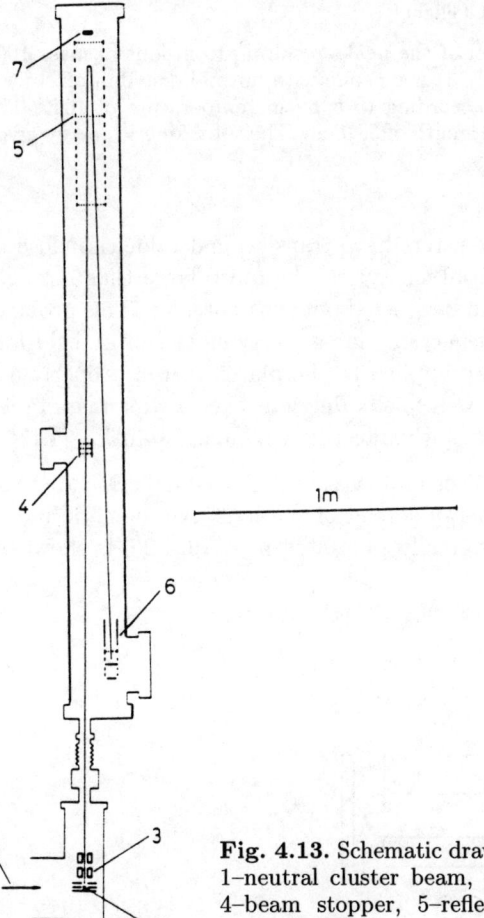

Fig. 4.13. Schematic drawing of the reflectron spectrometer: 1–neutral cluster beam, 2–ionization volume, 3–ion optics, 4–beam stopper, 5–reflector, 6–detector, 7–aux detector. (From [4.28])

Exact solutions of the equations describing the energy dependence of the ion flight time in a RTOF/MS can be obtained and the design parameters for a dual stage reflectron can be derived by analogy to what was described for a second order spatial focusing linear spectrometer [4.27]. With a standard RTOF/MS spectrometer a resolution of several thousands can be routinely obtained, the price to pay is a more complicate construction and a lower transmission.

Beergmann et al [4.28] have developed a RTOF/MS with a resolution $m/\Delta m$ of 35000, which is schematically shown in Fig. 4.13.

The remarkable performance of this spectrometer is obtained by a design and construction of the two-stage reflector assembly which guarantees very stable and homogeneous electric fields. Field homogeneity is a key factor for high resolution since field errors as small as 10^{-4} can cause significant time-of-flight errors. The reflectron stage corrects for the position spread due to the finite volume of the ionizing laser spot. The velocity spread is minimized by using the spectrometer perpendicular to the cluster beam, by the use of a large (6 kV) accelerating potential to turn the ions into the drift tube, and by ion optics consisting of two quadrupoles to correct the initial velocity components perpendicular to the spectrometer axis (Fig. 4.14).

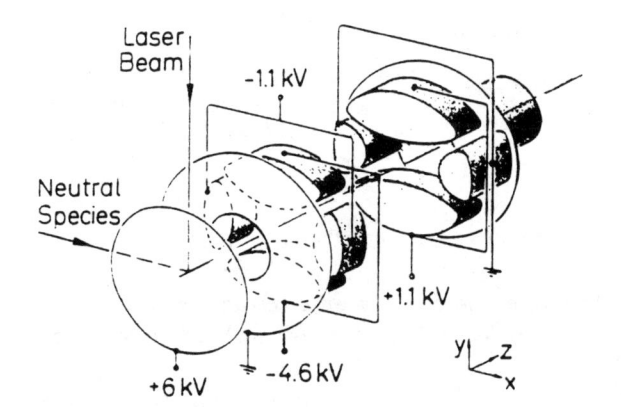

Fig. 4.14. Schematic drawing of the ion optics. (From [4.15])

4.1.3 Retarding Potential Mass Spectrometry

The determination of the mass of particles having the same velocity can be performed by analyzing their kinetic energies. Clusters in beams characterized by very high Mach numbers should have, in principle, substantially the same velocity and their kinetic energy should depend only on the number of their constituents. If the ionization of the aggregates does not change their energy, any device acting as an energy filter can behave as a mass spectrometer. The retarding potential mass spectrometer is based on this principle and

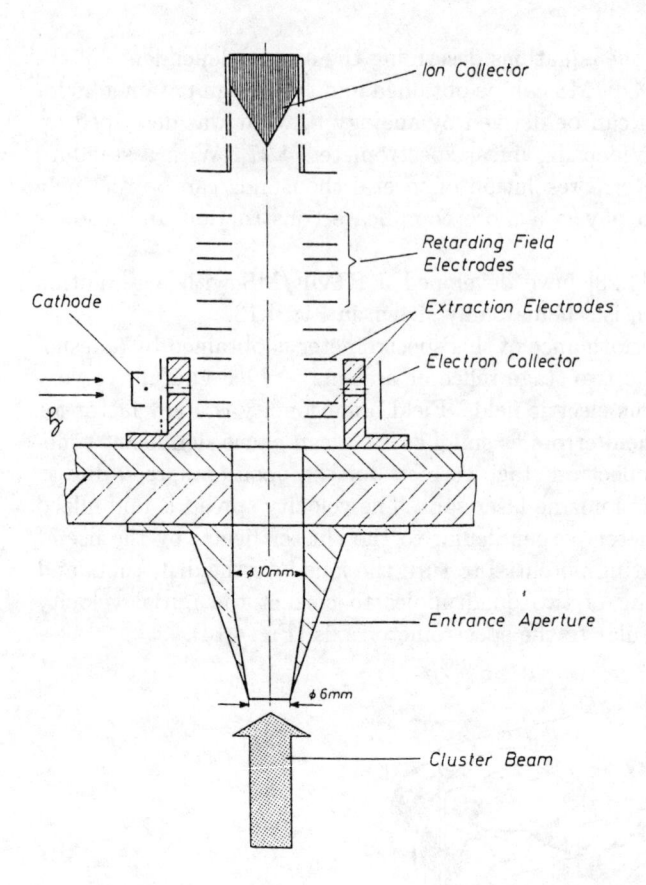

Fig. 4.15. Schematic drawing of retarding mass spectrometer for clusters. (From [4.30])

it is extensively used for the characterization of cluster beams (Fig. 4.15) [4.29, 4.30].

In this spectrometer clusters are ionized by an electron beam and then extracted into the retarding field electrode system parallel to the direction of the flow velocity. Only those clusters with N components which have a kinetic energy $E_{\text{kin}} = Nmv^2/2 \geq ZeU_{\text{r}}$, are able to overcome the positive potential barrier of the retarding field and reach the ion detector. The ion current distribution can be obtained by differentiating the transmitted ion current measured at the detector.

Information on the cluster size distribution can be obtained only if the results of the retarding field mass spectrometer are interpreted with independent measurements of the particle velocities. This is particularly important in the case of cluster beams where velocity slip and different residence times in the cluster source affect the cluster velocities in the beam. Attention should also be given to verify that the residence of the clusters in the ionization region does not affect their energy. Space charge effects in the ionizer can dominate the ion dynamics, making the interpretation of retarding field experiment questionable [4.31]. In particular, the use of retarding field methods for determining the mass distribution of clusters produced by pure expansion sources for ionized cluster beam deposition, cannot be used as an univocal evidence for the existence of large clusters in the beam (see next chapter) [4.31].

4.2 Detection Methods

4.2.1 Ionization of Clusters

The energy of ionized particles can be easily manipulated and measured by using electromagnetic fields, moreover, the charge carried by the particles can be efficiently detected: for these reasons neutral cluster beams have to be ionized. Neutral clusters can be ionized with electron or photon beams carrying energies larger than a critical value (appearance energy) [4.32]. Ion clusters are usually vibrationally excited immediately after ionization, hence decay reactions of the excited particles can produce dissociation leading to a strong modification of the initial neutral cluster distribution.

Photoionization processes and ionization potentials (IP) of small clusters have been extensively studied, showing a dependence upon the electronic and geometrical structure of the particles [4.33]. For clusters containing more than several hundreds atoms, these oscillations reduce in amplitude and the ionization potential evolves asymptotically towards the bulk work function (WF). In the case of simple metal clusters, this behavior can be described in a very simplified way by the relation:

$$\text{IP} = \text{WF} + \frac{e^2}{2R}, \qquad (4.33)$$

where R is the classical radius of the cluster [4.34].

Single-photon ionization of clusters with UV lasers (usually excimers) greatly reduces fragmentation, however, due to the extremely low duty-cycle, it cannot be used to produce intense ionized beams. Electron impact ionization is a viable way to produce ionized clusters, reaching efficiencies up to 80% [4.35]. Compared to photon ionization, the correlation between the cluster ion spectrum obtained with electron impact and the neutral cluster distribution is not straightforward, since fragmentation and multiple ionization are observed [4.30, 4.33].

In an electron ionizer for cluster beam electrons are emitted by metallic filaments and accelerated through a grid structure of the anode. The ionization region, where the beam passes, is inside the anode cage, as schematically shown in Fig. 4.16.

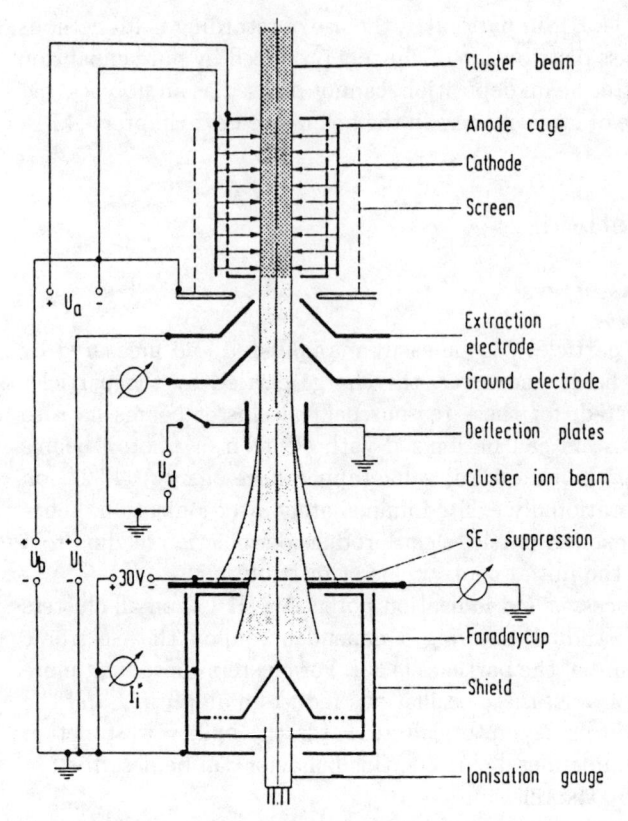

110   4. Characterization and Manipulation of Cluster Beams

Fig. 4.16. Experimental setup of the electron ionizer used for the ionization of hydrogen cluster beam. (From [4.35])

4.2.2 Charged Cluster Detection

Faraday Cup. A Faraday cup for positive ions consists of a cylindrical cup partially surrounded by a negatively polarized shield. This shield enhances the collection efficiency and prevents the detection of spurious electrons. The cup is connected to the input stage of an electrometer and is maintained at a virtual ground potential (Fig. 4.17) [4.5].

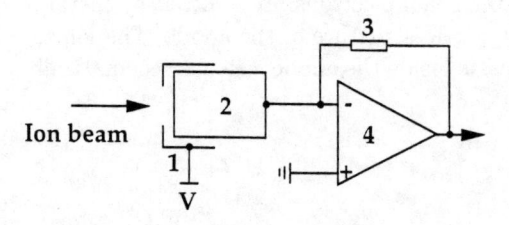

Fig. 4.17. Schematic representation of a Faraday cup. 1–electron repeller, 2–ion collector, 3–load resistor, 4–electrometer. (From [4.5])

This detector is very cheap and easy to build, and practical values of the load resistance R_L range from 10^8 and 10^{12} Ω: the signal to noise ratio increases with R_L. The load resistance should not be too high since the response time of the detector depends on R_L, moreover it should be always lower than the electrometer input impedance. With a careful minimization of noise sources a current as low as 10^{-15} A can be detected. This corresponds to an ionic flux of about 10^4 ions s^{-1}.

Electron Multipliers. Electron multipliers are a wide class of devices converting a positive ion current into an electron current which is further amplified by means of secondary electron emission effect. Their efficiency depends on the particle–electron conversion probability p. If F is the flux of particles impinging on the input stage of the detector, the output current I will be [4.5]:

$$I = -eFpG, \qquad (4.34)$$

where G is the gain factor which is typically in the range of 10^6 to 10^8. Electron multipliers can be used both in analog and pulse counting mode. In the analog mode, the electron current is sent to the first stage of an electrometer, where it is transformed into a voltage signal by means of a load resistor. To guarantee the linearity of response the electron current must be always lower than one tenth of the bias current of the electron multiplier. With this arrangement it is possible to achieve a signal which is G times the signal of a Faraday cup.

Pulse counting mode can be used when the ion flux is sufficiently low to enable the detection of electron bunches produced by individual ions. These pulses consist of G electrons distributed over about 10^{-8} s. The pulses are sent to the input stage of a preamplifier and converted into a voltage signal:

$$V(t) = \frac{eG \exp\left(\frac{-t}{R_I C}\right)}{C}, \qquad (4.35)$$

where R_I and C are the input resistance and the stray capacitance of the parallel combination of the multiplier and the preamplifier input stage. Capacitance should be reduced to increase the pulse height, The time constant $R_I C$ should be lower than 10^{-7} s to avoid pulse overlap.

Figure 4.18 schematically shows a discrete dynode electron multiplier consisting of an ion–electron conversion plate followed by a series of metallic dynodes.

Common multiplier materials are beryllium–copper (BeCu) and silver–magnesium (AgMg) alloys, covered by an active layer of composite oxides. For these surfaces p may range from 0.6 to 0.9 depending on the ion energy. Both p and G may be substantially reduced by modifications of surface composition caused by contamination.

A robust and economic alternative to discrete dynode multipliers are continuous dynode multipliers (channeltrons). They consist of a hollow glass tube covered internally by a semiconductor layer (Fig. 4.19)

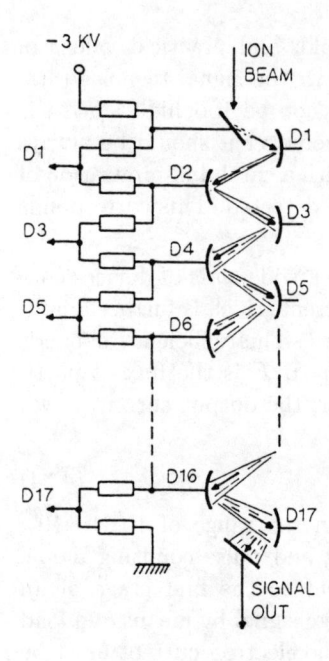

Fig. 4.18. Schematic view of a discrete dynode electron multiplier. Positive ions produce an electron current amplified by 17 dynodes. A resistive divider produces the multiplier polarization. (From [4.5])

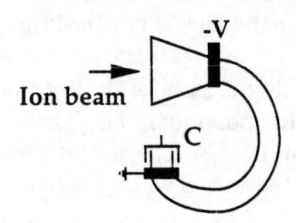

Fig. 4.19. Schematic view of a channeltron

The layer produces the secondary electrons and acts as resistor to establish the necessary voltage distribution along the multiplier. Ions hit the walls of the input aperture producing electrons which are accelerated along the axis of the tube. The gain of a channeltron depends primarily on the polarization voltage and on the ion flux. If the flux is below 10^4 ions s^{-1}, it depends only on the voltage.

Micro Channel Plates. A detector based on the same principle of continuous dynode multipliers and widely used for ion clusters detection is the micro channel plate (MCP). A MCP is an array of 10^4–10^7 miniature continuous electron multipliers oriented parallel to one another (Fig. 4.20).

Typical channel diameters are in the range of 10–100 μm and have a length to diameter ratio between 40 and 100 [4.36]. The channel matrix is usually fabricated from lead glass and the channel walls are covered with a semiconductor layer. Each channel can be considered as a continuous dynode, parallel electrical contact to each channel is provided by the deposition of

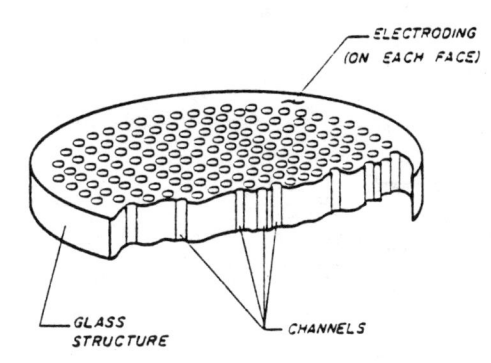

Fig. 4.20. Cutaway view of a MCP. (From [4.36])

a metallic coating on the front and on the rear surface of the MCP serving as input and output electrodes separated by a total resistance of $\sim 10^9 \, \Omega$. The MCPs, used singly or in cascade, allow a gain of 10^4–10^7 together with a time resolution considerably higher than that of other detectors (< 100 ps) and a spatial resolution limited only by the channel dimensions and spacing. The number of channels in a 25 mm diameter, 1 mm thick MCP with 25 µm diameter channels is about 5.5×10^5. Given the typical plate resistance, each channel has an associated resistance $R_c \simeq 2.75 \times 10^{14} \, \Omega$ and a capacitance of 3.7×10^{-16} F. Depending on the type of glass used, a dead-time of several milliseconds must be expected, however due to the large number of channels operating independently, a dead-time of 10^{-7}–10^{-8} s is observed, provided that a single channel is not excited more frequently than 10^{-2} s.

Straight channel MCPs usually operate at gains of 10^3–10^5, the upper limit being set by the onset of ion feedback. The probability of producing positive ions in the high charge density region at the output of the channel increases with the gain. The ions are produced by electron collisions with residual gas molecules at pressures greater than 10^{-6} torr and with molecules desorbed from the channel walls under electron bombardment. These ions can drift back to the channel input, producing ion after pulses [4.36].

Ion feedback suppression can be achieved by using two MCPs in series in the so-called Chevron configuration (Fig. 4.21).

The MCPs are oriented so that the channel bias angle provides a sufficiently large bias angle to inhibit backdrift of secondary positive ions from the rear plate back to the front plate. A gain curve of a single MCP and for a Chevron is reported in Fig. 4.22. Typically the different MCPs are separated by 50–150 µm and operated at a gain of 10^4.

The detection efficiency of MCPs to various kind of primary radiation is summarized in Table 4.1.

The performance characteristics of MCP have been extensively tested for nuclear physics applications, the results can be extrapolated to clusters using great care since cluster–surface interaction causing charged particle emission presents several differences compared to ion–surface interaction.

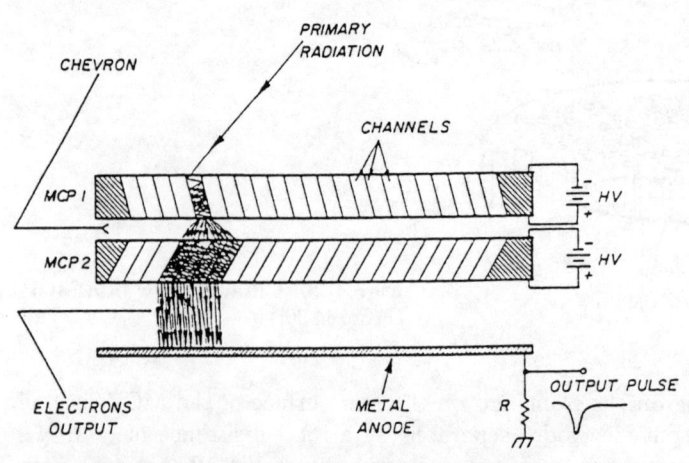

Fig. 4.21. Two MCP in chevron configuration (side view). A conical 50 Ω anode is used to collect the charges coming from the rear surface of the second MCP. (From [4.36])

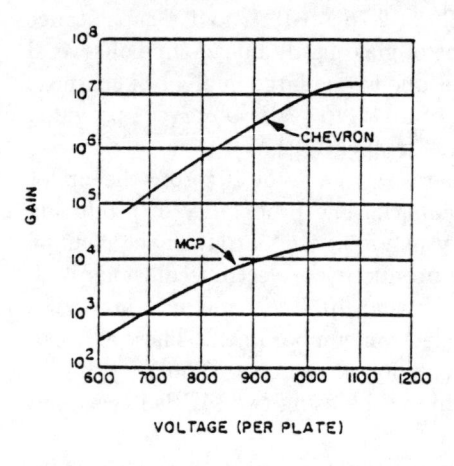

Fig. 4.22. Gain vs voltage characteristic for a straight channel MCP and a Chevron. The input current signal is $1 \times 10^{-12}\,\text{A cm}^{-2}$ and the input energy 300 eV. (From [4.36])

Table 4.1. Detection efficiency of MCP. (From [4.36])

Type of radiation		Detection efficiency %
Electrons	0.2–2 keV	50–85
	2–50 keV	10–60
Positive ions	0.5–2 keV	5–85
	2–50 keV	60–85
	50–200 keV	4–60
UV radiation	30–110 nm	5–15
	110–150 nm	1–5

The interaction of ions with surfaces and the subsequent secondary electron emission is a well studied phenomenon [4.37]. The yield of secondary electron emission from an ion impact on a target decreases with decreasing ion velocity down to a threshold velocity: this value has been found to be around $2 \times 10^4 \, \mathrm{m\,s^{-1}}$ depending slightly on ion and target material [4.37].

The interaction of clusters, especially in the large size range, and surfaces is considerably less studied and understood, although this is a fundamental issue for the detection of large clusters and other systems such as biomolecules, via secondary electron emission [4.38–4.40].

It is very important to determine the velocity threshold for secondary electron emission from cluster impact. In fact, using a constant acceleration potential for cluster–detector impact, clusters with increasing mass will have decreasing velocities.

The emission of charged particles upon impact, is not limited to ions at hypervelocities, but it can occur also for clusters with velocities as low as 1–$2 \, \mathrm{km\,s^{-1}}$ and also from the impact of neutral aggregates [4.41,4.42]. Even et al [4.41] report measurements of electron emission from neutral CCl_4 clusters at $1600 \, \mathrm{m\,s^{-1}}$ in a mass range which extends up to about 1000 amu. Water clusters, with an average size of 2700 molecules, and at an impact speed of $1300 \, \mathrm{m\,s^{-1}}$, emit ionized cluster fragments when they hit a clean graphite surface, heated at 950 K. In the case of surfaces that have been exposed to air at atmospheric pressure the emission occurs also at room temperature [4.42].

The behavior of MCPs with very large mass clusters is basically unknown. Martin and co-workers have shown that very large alkali clusters (up to several 10^5 amu) can be detected with MCP [4.43]. Figure 4.23, shows the relative probability of detection of cesium atoms: small clusters are detected independently from their energy (section I), clusters in section II are detected independently from their mass and depending on their energies. The detectability of large clusters in section III presents a complex dependence on the cluster energy and on the energy per atom.

Scintillator Detector. The secondary electron emission produced by the impact of energetic particles on oxide films can be exploited to construct a simple, robust and efficient detector for ionized clusters. The so-called scintillation type ion detector has been proposed by Daly [4.44].

Charged particles entering the detector are accelerated at high energy (several tens of kV) against the surface of a collector plate thus producing secondary electrons. These, in turn, are accelerated in the same field and strike a plastic scintillator, giving rise to a light signal detected by a photomultiplier mounted externally to the vacuum chamber. Several configuration have been adopted for ion detection, showing performances comparable to those of MCP-based detectors [4.45, 4.46].

The Daly detector offers several advantages: high secondary emission yield, large detector area, it is cheap, robust, compatible with HV and UHV systems, easy to construct and it has a very low dark noise. This detector can

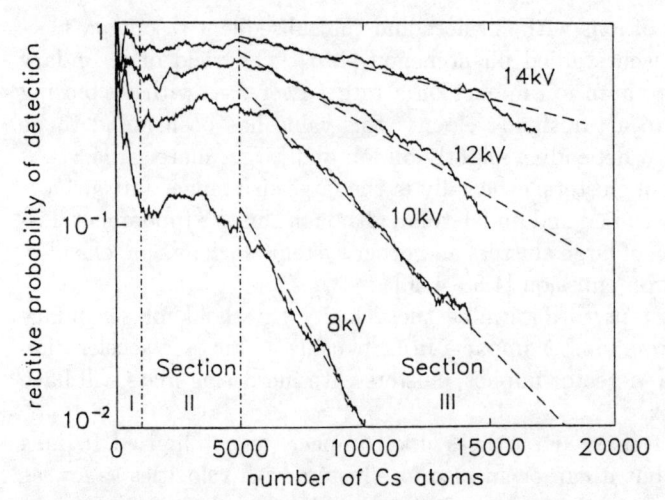

116   4. Characterization and Manipulation of Cluster Beams

Fig. 4.23. Relative probability of detection of cesium clusters. The signal is normalized to the signal of maximum acceleration voltage (16 kV). The size range can be divided into three sections: a saturation regime (**section I**), a regime of constant relative detection probability (**section II**), and a regime of exponential decay (**section III**). The interpolating lines in section III have a slope of -8 eV/E_{atom}. (From [4.43])

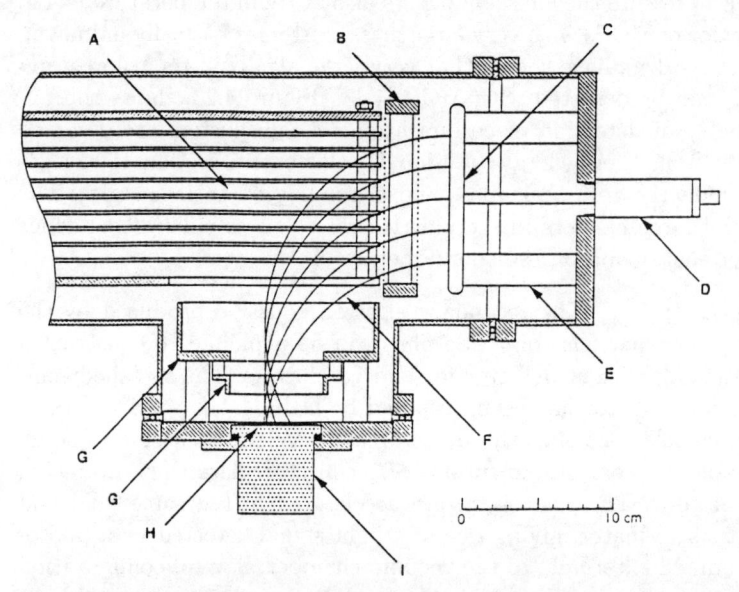

Fig. 4.24. Horizontal cross-section of the ion cluster detector. A: field-free region, B: grid, C: Al coated dynode, D: high voltage feedthrough, E: ceramic standoff, F: secondary electron trajectory in the perpendicular magnetic field, G: electrostatic lenses assembly (grounded), H: scintillator, I: Lucite light pipe. The photomultiplier (not shown) is at the end of the light pipe. (From [4.17])

be efficiently used coupled to a TOF/MS in a cluster beam apparatus [4.17]. A cross-section of a Daly detector used in clusters experiments is shown in Fig. 4.24.

Positive ions which pass through the end grid of the mass spectrometer field-free region, are accelerated against a dynode kept at high negative potential. The emitted electrons are focused on the scintillator by a uniform perpendicular magnetic field produced by two Helmoltz coils mounted outside the detector chamber. Photons are detected by a fast photomultiplier coupled to the scintillator by a Lucite light pipe. A mass spectrum of aluminum clusters obtained with this detector is shown in Fig. 4.25

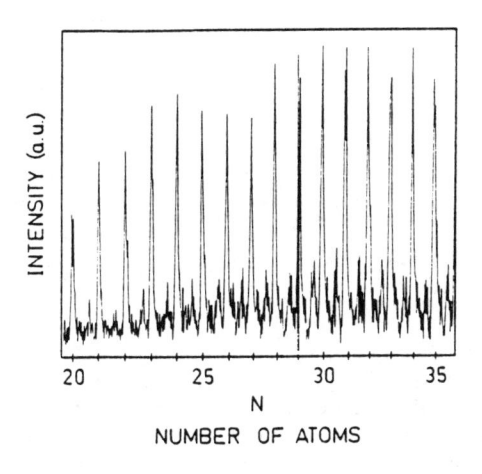

Fig. 4.25. Mass spectrum of Al clusters produced by a single shot of a laser vaporization source. The detector is placed about 2 m away from the source. Vertical full scale corresponds approximately to 150 cluster ions. (From [4.17])

4.2.3 Cluster Beam Characterization

Due to the extensive use of pulsed and modulated sources for cluster production, it is important to characterize the cluster beams in terms of time profile and velocity distribution.

The velocity distribution of clusters is a key parameter for cluster deposition experiments and strongly affects the resolution of TOF/MS. A precise characterization of the cluster velocity distribution can be performed using TOF methods [4.47, 4.48].

A TOF measurement consists of the determination of the pulsed beam flux intensity at two different points of its trajectory. In the case of continuous beams or long smooth pulses, a chopper should be used for beam modulation [4.48, 4.49]. The beam intensity profile at the second detection point is simply given by the evolution of the profile at the first detection point, caused by the velocity distribution of the species contained in the beam (Fig. 4.26).

With a chopper, one obtains, at the first point, a narrow rectangular pulse from the molecular beam. The velocity distribution is thus determined

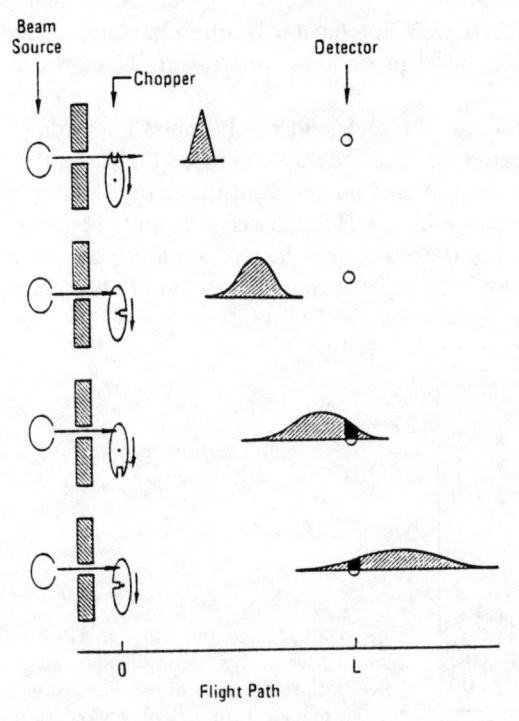

Fig. 4.26. Schematic illustration of a TOF measurement. The evolution of the molecular beam modulated by the rotating disk chopper is shown during the travel towards the detector. The TOF signal at the interception of the detector is the black portion of the molecular beam density distribution. (From [4.48])

directly from the shape of the pulse at the second point spread out by the effect of different velocities. Alternatively, one can measure a structured pulsed-beam intensity profile at the two detection points and subsequently extract the velocity distribution by a fitting procedure, based on the convolution of the first pulse with a suitable distribution function [4.50]. In this case the principal features of the beam profile must have a characteristic time scale, longer than that associated with detector response.

The response of any fast beam-intensity detector can be used to determine velocity distributions of molecules in a pulsed beam. Both flux-sensitive and density-sensitive beam detectors [4.48] can be used: a typical detector of the latter type is the fast ionization gauge (FIG) [4.47]. This instrument is a simple and effective monitor of the molecular beam and it consists of a standard Bayard–Alpert gauge tube with a central ion collector wire surrounded by a helical grid and with the electron-emitting filament on the outside [4.47]. The ion current in the FIG is proportional to the electron emission and to the istantaneous gas density inside the grid. The fast response (μs) of the detector is obtained by having the ion collector wire connected directly to the negative input of an operational amplifier in order to minimize the input capacitance. To obtain this, the amplifier circuit must be built into the gauge head and operated under vacuum.

This very compact detector can be easily displaced in a vacuum apparatus and used to perform TOF measurements of the velocity distribution of pulsed beams. In particular, the operation of a pulsed valve (shutter function, pulse width, etc.) can be characterized.

In the case of cluster beams, i.e. seeded supersonic beams, the FIG cannot be used since it cannot discriminate between the highly diluted component of the beam and the carrier gas. This is particularly true for pulsed plasma cluster sources, where there is often a huge mass difference between the carrier gas and the diluted species because of condensation, so that a high dilution ratio has to be maintained to limit the velocity slip. This is not always possible because the amount of carrier gas is restricted to fixed boundaries, limited by source operation range. It is thus reasonable to expect that cluster kinetics may be considerably different from that of the carrier gas.

A channeltron can be used to characterize the velocity distribution of neutral clusters. Typically cluster sources produce beams at velocities of roughly $2000 \, \mathrm{m\,s^{-1}}$. Supposing a moderate velocity slip, the kinetic energy of the clusters if of several hundreds of eV. At these energies, neutral particles impinging on a channeltron can cause the production of charged species upon impact, the resulting channeltron signal can be conveniently used for TOF cluster beam characterization.

The beam profile (Fig. 4.27) can be obtained by sampling the channeltron signal with a digital oscilloscope: the voltage measured by every oscilloscope channel can be considered proportional to the number of emission events in the corresponding time interval, and hence, scaled by detection efficiency, to the beam flux intensity. However, this proportionality holds only provided that the time interval between the detection of two subsequent impact events is much longer than the channeltron dead time. When operating in the range of non linear response of the "counting" device [4.51], the channeltron signal is still a measurement of the rate of emission events, but must be multiplied by an intensity-dependent scaling factor, in order to compensate for losses in channeltron gain. Figure 4.27 shows an intensity profile of a carbon cluster beam obtained with this technique. The estimated peak of emission events rate is $6 \times 10^6 \, \mathrm{s^{-1}}$, with an exponentially decreasing tail resembling the gas pressure fall in the cluster source [4.50].

The detection efficiency of a channeltron can be evaluated by independent cluster flux determination. From quartz microbalance growth rate measurements, the channeltron detection efficiency in the above example has been determined to be $\sim 5 \times 10^{-8}$. Of course, this efficiency has to be considered as a weighted average of single efficiencies over all the masses in the beam, the weight being the cluster mass distribution. For a channeltron it is reasonable to expect a size dependence of the efficiency similar to that reported in [4.43] for MCP.

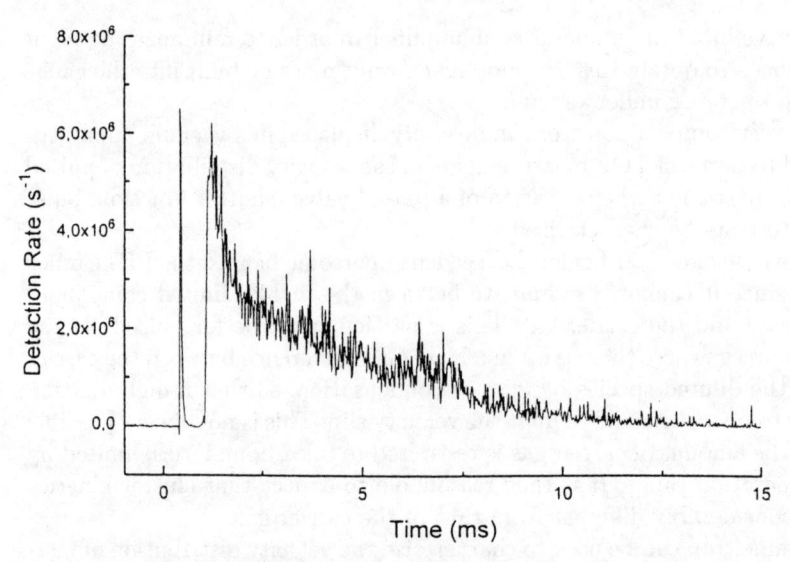

Fig. 4.27. A typical intensity profile obtained as described in the text. The signal is terminated at the low impedance (50 Ω) input of a digital oscilloscope. The electron multiplier gain is *8 × 10⁷* at 2500 V supply voltage, when the counting rate is lower than 105 Hz. For higher rates different gain values have been measured and used for profile calibration. The sharp peak preceding the cluster pulse is due to ultraviolet light from the spark directly illuminating the channeltron

4.3 Cluster Selection and Manipulation

4.3.1 Size and Energy Selection

Mechanical Selection. As we discussed previously, a rotating disk chopper can conveniently modulate a cluster beam to perform TOF measurements. If clusters are produced in a pulsed mode and if the length of the pulse is smaller than the chopper period, this method can be used also for mass selection [4.52]. If the velocity distribution of the clusters is known, by a proper synchronization of the particle exit time from the source with the moment when it reaches the chopper, it is possible to select the desired mass portion of the cluster pulse. Deposition rates of $7\,\mathrm{nm\,cm^{-2}min}$ have been reported, using this method, for Si clusters produced by laser-induced decomposition of SiH_4 in a flow reactor [4.52, 4.53]. A major advantage of mechanical selection is that one does not need ionized cluster beams.

Wien Filter. In a Wien velocity filter, mass separation is achieved with crossed homogeneous electric field E and magnetic field B, perpendicular to the ionized cluster beam. Figure 4.28 shows a schematic representation of a Wien filter.

Assuming B in the x-direction and E in the y-direction, the equation of motion of a particle with mass m and charge q are:

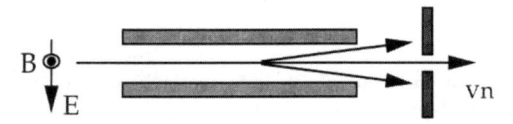

Fig. 4.28. Schematic representation of a Wien velocity filter

$$m\frac{dv_x}{dt} = 0, \qquad (4.36)$$

$$m\frac{dv_y}{dt} = q(E + v_z B), \qquad (4.37)$$

$$m\frac{dv_z}{dt} = -qv_y B. \qquad (4.38)$$

With a homogeneous field the equations are exactly solvable:

$$x = v_{x,0} + x_0 \qquad (4.39)$$

$$y = \frac{m}{qB}\left((E/B + v_{z,0})(1 - \cos(qBt/m)) + v_{y,0}\sin(qBt/m)\right) + y_0 \quad (4.40)$$

$$y = \frac{m}{qB}\left((E/B + v_{z,0})\sin(qBt/m) + v_{y,0}(\cos(qBt/m) - 1)\right)$$
$$- Et/B + y_0. \qquad (4.41)$$

These solutions show that the particles move on cycloidal trajectories, the deflection of particles deviating for a certain fraction from the nominal velocity $v_n = E/B$ is a measure of the resolving power of the filter. The determination of mass and velocity distributions of particles in a beam can thus be performed by selecting with the filter a particular velocity and then using a retarding potential to determine the mass [4.54]. It is, of course, very important to know the initial particle velocity distribution, in order to make precise measurements. As already discussed, the velocity distribution of clusters in beams is affected by several factors such as velocity slip and it is, in general, a function of the mass of the aggregates. Moreover, it should be remembered that the use of cluster sources with thermalization cavities make the velocity distribution very complex and difficult to predict. To exploit the advantages of Wien filters (high transmittance, easy operation and construction) compared to other methods for cluster mass selection, one has to reduce the initial particle velocities and/or determine this distribution with precision. Size selection of very small metallic clusters (few tens of atoms) has been performed with a Wien filter coupled to a PACIS source [4.55]. The initial kinetic energy of the clusters is measured with a retarding potential method [4.56].

4.3.2 Quadrupole Filter

The deposition of cluster beams monochromatic in size and energy is, in principle, the ultimate goal for the synthesis of nanocrystalline materials with tailored properties. Several experimental problems remain to be solved before size-selected deposition becomes viable for the synthesis of nanostructured materials. In particular, the development of both high intensity sources of ionized clusters and mass/energy selection systems with a high transmission are necessary prerequisites. Despite the deposition of size-selected clusters in bulk quantities being far from achieved, controlled deposition of clusters to study cluster-surface and cluster–cluster interactions in the sub-monolayer regime and cluster catalytic properties have been realized.

Cluster size selection requires high fluxes of ionized particles. The nominal current densities should be higher than $1\,\mathrm{nA\,cm^{-2}}$ in order to deposit a consistent fraction of a monolayer in a time reasonably short to prevent contaminations. In a UHV regime, with pressures of the order of 10^{-11} torr, depositions should take typically less than one hour (with a current of $1\,\mathrm{nA\,cm^{-2}}$, this means a deposit of roughly 10% of a monolayer) [4.57].

High quantities of very small ionized clusters can be produced by sputtering a target with fast atom or ion beams. In order to obtain a high sputtering yield heavy ions at high energies and fluxes must be used. One of the most popular ion sources used in this type of experiment is the CORDIS source capable of delivering Xe^+ currents of 10 mA at a beam divergence of 20 mrad [4.57]. This source can be coupled to a size-selection device based on a quadrupole spectrometer [4.58].

Quadrupole mass spectrometers are used as mass filters and ion guides for cluster deposition. They have a good mass resolution in the small size range and for particle kinetic energies of a few eV. This is the typical energy range of particles produced by sputtering and it is a good range for soft-landing [4.59]. A scheme of an experimental set up for size-selected cluster deposition is shown in Fig. 4.29

Deposition data for Pt clusters obtained from this kind of apparatus are shown in Table 4.2.

Table 4.2. Deposition data for Pt monomer and clusters. (From [4.57])

Pt_n	Ave. current ($nA\,cm^{-2}$)	Deposition time (min)	Surface atom density ($10^{14}\,cm^{-2}$)
1	60.0	6.0	1.35
2	20.8	7.5	1.17
3	5.10	23.0	1.3
10	0.35	98.0	1.29

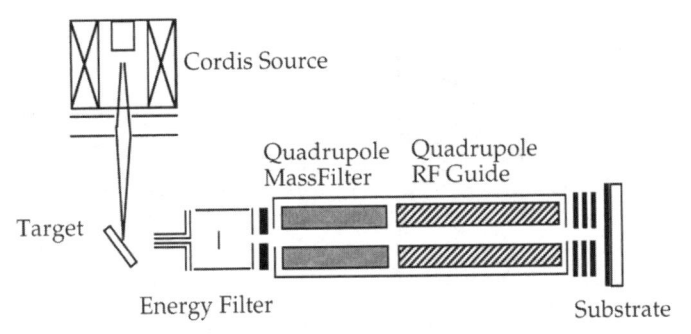

Fig. 4.29. Schematic of an experimental set up for production, deposition and *in situ* characterization of size-selected clusters. (From [4.57])

The intensity of CORDIS sources decays exponentially from the monomer ion to larger clusters, while a laser vaporization source has been demonstrated to be suitable for the production of reasonably high currents of clusters up to twenty monomers [4.60]. A critical parameter for controlled deposition is the kinetic energy of the clusters prior to mass selection and at the deposition. Experiments have shown that massive fragmentation of clusters on surfaces is prevented if clusters are landed with energies of few eV per atom [4.61, 4.62]. To this purpose, clusters are decelerated by a grid system prior to land on the substrate. A charge cloud of electrons is maintained between the deceleration grid and the substrate to neutralize the aggregates [4.63].

4.3.3 Separation of Gas Mixtures in Supersonic Beams

As already discussed at the end of Sect. 2.1.2, separation effects consisting of an enrichment of heavy species in the center of the beam are observed in seeded supersonic expansions [4.64, 4.65]. To account for these phenomena Reis and Fenn [4.65] proposed an analogy between seeded beams and aerosol beams [4.66]. As shown schematically in Fig. 4.30, the deceleration field experienced by axial streamlines between the onset of the shock and stagnation and the skimmer can divert light species from the center of the beam: inertial impactors for the separation of particles in aerosols are based on this principle [4.65–4.67].

The analogy with aerosols should hold even better in the case of supersonic cluster beams. Experiments conducted on carbon and NiTi cluster beams produced by a pulsed discharge source, show the possibility of selecting different mass distributions by sampling different regions of the beam spot [4.68, 4.69].

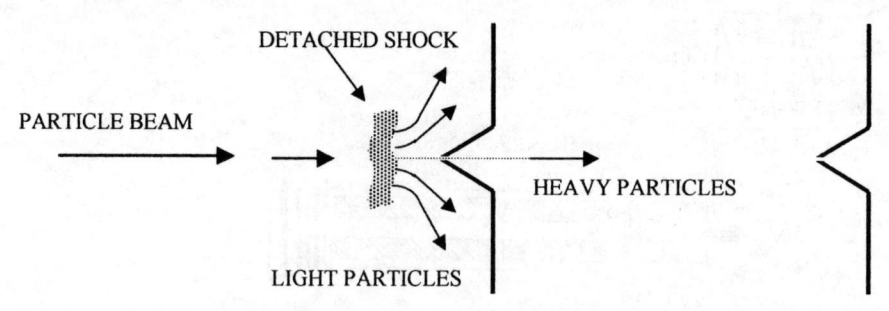

Fig. 4.30. Schematic illustration of flow field behind bow shock on skimmer

5. Thin Film Deposition and Surface Modification by Cluster Beams

The bombardment of solids by ions or energetic neutrals gives rise to a variety of phenomena widely utilized for the modification of surfaces and the deposition of thin films with tailored properties [5.1]. Depending on ion energies and fluxes, ion beams induce etching, sputtering, implantation, and thin film growth. These effects are schematically shown in Fig. 5.1.

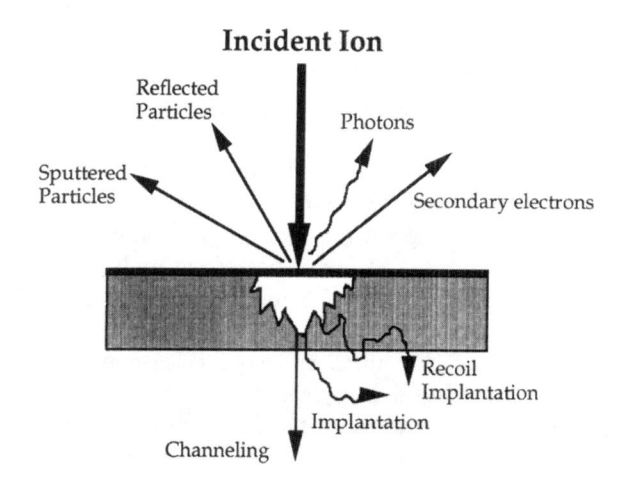

Fig. 5.1. Schematic diagram of ion surface interaction

Cluster beams are emerging as a powerful and versatile tool for the modification and processing of surfaces as an alternative to ion techniques. This trend is driven by their interesting characteristics, such as high sputtering and etching rates, very low kinetic energy per atom and extremely high density bombardment effects (Fig. 5.2) [5.2].

Cluster ion beams are proposed to overcome the limitations of ion treatments for anisotropic etching and nanofabrication technology [5.3]. Compared to ions, cluster beam impact generates multicollision effects in an energy range from a few eV up to a few hundred eV per atom at high densities. These characteristics are well suited for obtaining very shallow implantation, high yield sputtering, surface cleaning and structuring. At typical cluster kinetic energies of a few tens of keV each atom carries a very low energy minimizing damage. Since the charge to mass ratio is very favorable, cluster

126   5. Thin Film Deposition and Surface Modification by Cluster Beams

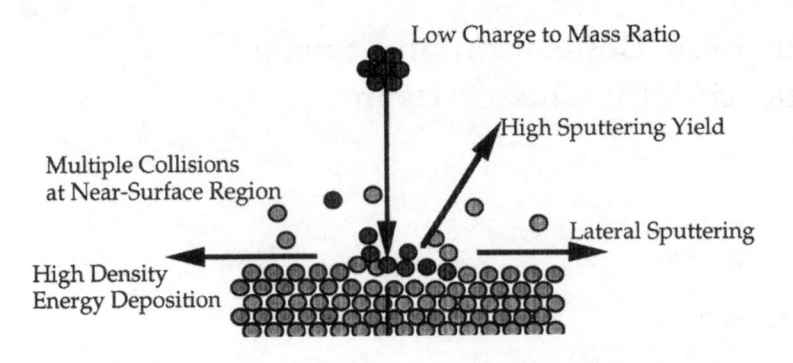

Fig. 5.2. Characteristics of cluster ion beam processes. (From [5.2])

impact allows the deposition of high density energy localized in the surface region, minimizing implantation and channeling. Moreover, cluster beams provide high directionality useful for anisotropic etching and it can prevent the damage to the device caused by accumulation of charge formed by charged species impinging on insulating substrates as observed in plasma etching of semiconductor devices [5.4].

It has been demonstrated that ion bombardment during deposition of thin films strongly influences nucleation and growth kinetics by enhancing the adatom surface mobility (Fig. 5.3), however inclusion of the bombarding species in the film represents a problem [5.1]. Cluster beams can reduce the amount of surface damage and implantation while keeping the same beneficial characteristics of ion beams.

Cluster beam deposition (CBD) of thin films and composite materials can be considered as a technique with features typical of different synthetic methods such as atom and ion beams on the one hand and plasma spray deposition on the other. Cluster beams offer the possibility of controlling the kinetic energy of the particles as for ion beams, moreover, the mass of

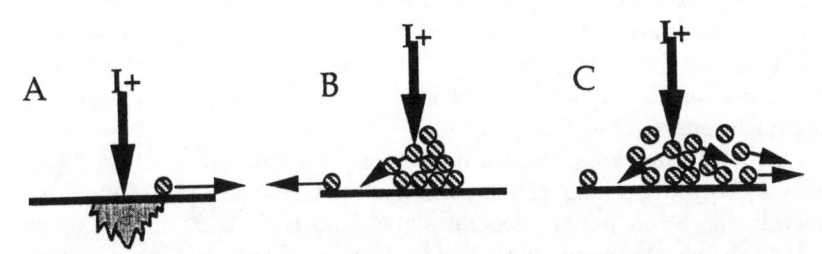

Fig. 5.3. Schematic illustration of three possible mechanisms of ion enhanced surface diffusion. (**A**) Direct coupling of energy to single adatoms through the ion generated phonon cascade. (**B**) Forward sputtering from edges of three-dimensional islands. (**C**) Ion induced dissociation of small clusters. (Adapted from [5.1])

projectiles can be varied in order to obtain, by cluster assembling, a new class of solids which lie between disordered and crystalline.

Another peculiar aspect of CBD is that the conversion of kinetic energy of a cluster colliding with a surface initiates reactions and coalescence phenomena characterized by conditions difficult to obtain with monomer beams. The degree of coalescence and particle–substrate adhesion depends on the kinetic energy. By varying this parameter, one can obtain a variety of different structures and morphologies ranging from granular thin film retaining the structures of the precursor clusters to strongly adhering smooth and compact thin films.

Historically, the use of cluster beams has been proposed for both very low-energy and high kinetic energy regimes. The control of packing phenomena of ultrafine powder having a diameter below 0.1 µm is very important for material processing application but also very difficult from the point of view of powder handling. Cluster beams represent an interesting solution as reported, in a pioneering work, by Kashu et al [5.5] who proposed to accelerate towards a substrate small particles produced by gas aggregation. Particles having diameters in the range of several tens of nanometers, acquire velocities on the order of $100 \, \text{m sec}^{-1}$ corresponding to kinetic energies that typically produces coalescence of particles in the collision zone. This phenomenon is usually called ballistic compaction.

The acceleration of ionized clusters and their impact at high energies (keV) on substrates to synthesize high quality thin films and heterostructures has been proposed by Takagi and Yamada [5.6]. Several groups have adopted it for depositing thin films of various materials including semiconductors, metals, dielectrics, magnetic and organic materials.

In between the mentioned two extremes, there is a largely unknown realm of cluster-assembled materials, the exploration of which is only at the beginning.

In view of a systematic characterization, at least qualitative, of the different regimes and applications of CBD, we distinguish cluster beam interaction with static (non-growing) surfaces and with dynamic (growing) surfaces. On static surfaces, where ion bombardment is used for preparing atomically clean surfaces [5.7] or to produce sputtering, cluster beams can represent an alternative for surface cleaning, and etching, the advantages being a higher sputtering yield and a low degree of sub-surface induced defects. In the case of growing surfaces, thin films with a wide variety of structures and morphologies can be synthesized, depending on cluster beam kinetic energy, mass distribution and intensity.

5.1 Kinetic Energy Regimes

Although the microscopic mechanisms underlying cluster beam deposition are still awaiting a precise experimental characterization, many molecular dynamics (MD) simulations have modeled this process with the aim of correlating the structure of the film with the kinetic energy of the deposited clusters. Due to the limited temporal window available, MD simulations can describe only the initial steps of the cluster–surface interaction and evolution, so that the fate of the cluster fragments on the surface, after the impact, cannot be followed. It should also be recognized that several parameters of the cluster beam and of the cluster–surface interaction (mass distribution and deposition rate of the clusters, cluster structure and temperature, cluster structure on the surface, etc.) are poorly known, moreover, the mass range accessible to MD simulations is often very different from that used in experiments. Keeping in mind these limitations, MD simulations can provide an indication of the relevant processes underlying cluster–surface and cluster–cluster interactions and give a qualitative picture of the stages of the evolution of cluster-assembled structures in a region of times and dimensions hardly accessible to experimental probes.

The time evolution of thin film growth by cluster beam deposition has been studied using a two-dimensional molecular dynamics simulation [5.8]. It is assumed that one atom interacts with another via a pairwise additive and spherically symmetric potential having the form:

$$V(r) = 4\varepsilon[(\sigma/r)^{12} - (\sigma/r)^6], \tag{5.1}$$

where the units of length and energy are assumed to be σ and ε, respectively. The time-dependent local heating, atomic rearrangements, recrystallization and coalescence of clusters impinging on a perfect surface has been investigated at different cluster kinetic energies ($E/N = 0.1, 1, 2$ and 4ε) for a 91 atom cluster (Fig. 5.4). Assuming the order of magnitude of the cohesive energy of a metallic cluster, 0.1ε corresponds to an energy typical of particle ejected from a gas-aggregation source. A kinetic energy increase causes a deformation and breaking of the cluster, while at high energy damage of the surface occurs without annealing.

No significant enhancement of surface atom mobility in terms of diffusion along the surface is observed in any of the cases studied. Figure 5.5 shows the evolution of the microstructure of a film obtained by the deposition of 13 clusters on a surface at different energies. At the lowest kinetic energy the clusters remain almost intact forming a film consisting of crystallites with dimensions of the original clusters. At higher energies the clusters start to coalesce and recrystallization occurs although the film presents a consistent fraction of voids. Increasing the energy to the atomic bond strength, homoepitaxial growth is observed. The packing density increases with cluster kinetic energy. When the cluster energy exceeds $E/N > 4\varepsilon$ sputtering of the surface and defect creation is observed.

5.1 Kinetic Energy Regimes   129

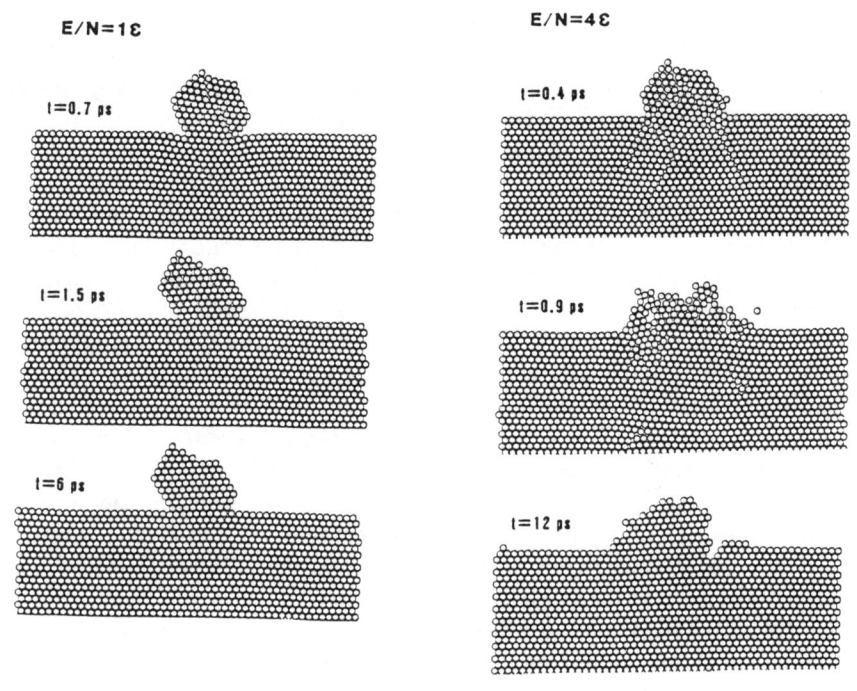

Fig. 5.4. Cluster hitting a perfect substrate at different initial kinetic energies. (From [5.8])

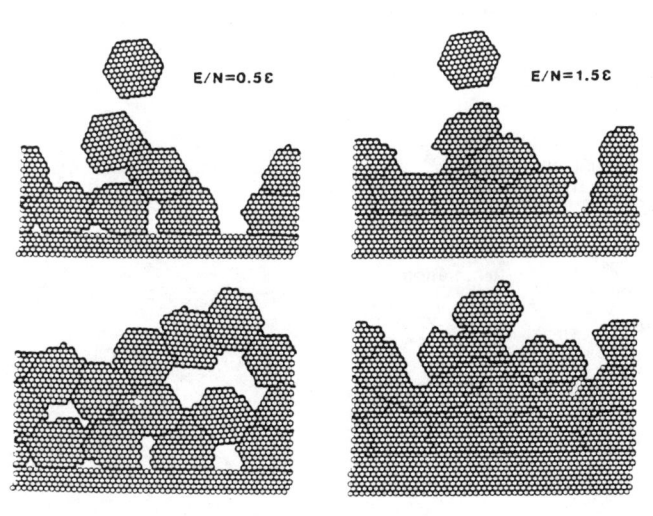

Fig. 5.5. Film microstructure evolving when several clusters successively approach the substrate. (From [5.8])

Cleveland and Landman [5.9] performed a MD simulation of the collision of an Ar cluster formed by 561 monomers with a (001) NaCl surface at a kinetic energy of $3\,\mathrm{km\,s^{-1}}$ (1.863 eV per atom). The impact of the cluster on the surface results in a pileup phenomenon leading to high energy collision cascades and the development in the cluster of extremely high density, pressure and temperature ($\geq 10\,\mathrm{GPa}$, 4000 K). The solid undergoes severe deformation and disordering in a small region localized around the impact area of the cluster and extending several layers beneath the surface. Implantation of argon atoms in the disordered surface is also observed. For a cluster velocity of $10\,\mathrm{km\,s^{-1}}$ massive surface damage and cluster penetration occurs.

MD simulations of cluster deposition on silicon have shown that epitaxial growth can be achieved only in the presence of a high surface diffusion that is attainable with large initial cluster velocities and moderate substrate temperatures [5.10,5.11]. Small amorphous silicon clusters (33 atoms) were deposited at various translational energies ranging from 0.35 eV to 2.1 eV on Si (111) surfaces. The energy deposition and spreading of the atoms of the clusters are characterized as a function of surface temperature and cluster temperature and size, showing that by using kinetic energies higher than those of conventional processes allowing the use of reduced substrate temperatures while keeping a high surface diffusivity.

The interaction between energetic clusters and metallic substrates have been simulated by molecular dynamics for Cu, Ni and Al clusters impinging on the same substrate in an energy range between 92 eV and 1 keV [5.12]. The simulations are based on an embedded atom method potential and show that elastic and chemical properties of the cluster–surface combination, as well as the relative mass of the cluster and substrate atoms strongly influence the deposition features. These findings are summarized in Fig. 5.6 where

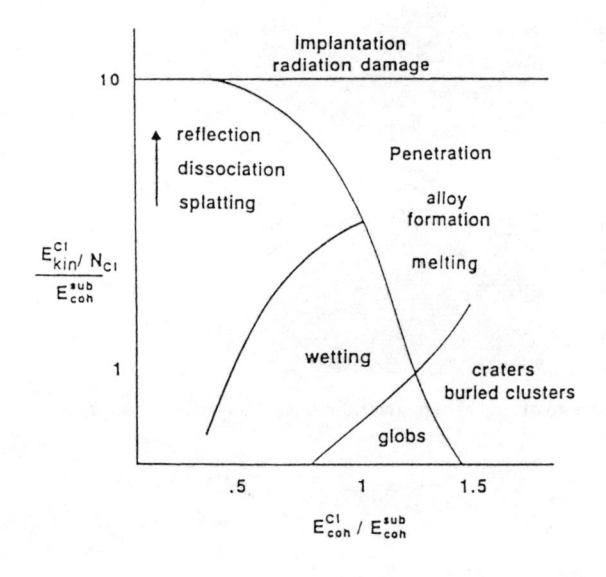

Fig. 5.6. Mechanism diagram of the interaction between energetic clusters of atoms and metal substrates. (From [5.12])

a diagram describing different interaction regimes is presented. Besides the specific cases that are, however, hardly comparable with the available experiments, a main observation is assessed: the actual cluster behavior, in the regime where a tendency to form compact films may arise, is quite complex since the impact, the heat transfer and the atomic vibrations, all occur on similar time scales.

Langevin-molecular dynamics simulations have been used to characterize the impact of metallic clusters on a metallic substrate (Mo_{1043} on Mo (001) surface) with impact energies of 0.1, 1, 10 eV atom^{-1} [5.13]. This system has been modeled by an embedded atom potential. The effect of a single cluster impact is shown in Fig. 5.7.

The cluster is completely destroyed for an impact energy of 10 eV and a strong shock wave is generated. A peak pressure of 100 GPa is produced during the impact and no ejection of atoms is observed. Different film morphologies are produced by different kinetic energies, as shown in Fig. 5.8.

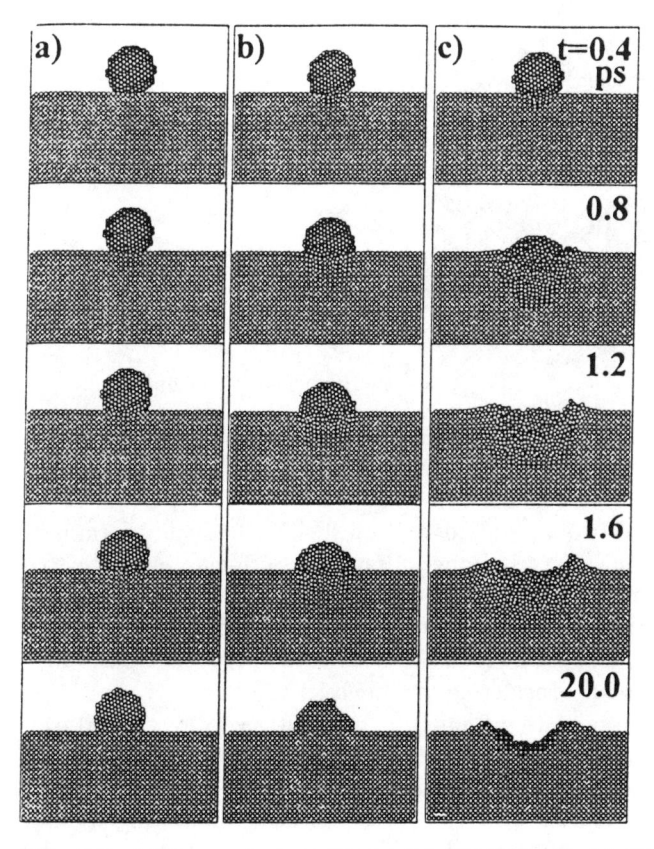

Fig. 5.7. A Mo_{1043} cluster with (a) 0.1 eV, (b) 1 eV, (c) 10 eV kinetic energy per atom impinging on a Mo(001) surface. (From [5.13])

Fig. 5.8. Morphology of a film formed by Mo_{1043} cluster with (a) 0.1 eV, (b) 1 eV, (c) 10 eV kinetic energy per atom. (From [5.13])

Deposition at 10 eV atom^{-1} ends with a dense epitaxial film well adhered to the substrate.

Slow clusters (0.1 eV atom^{-1}) almost retain their individuality forming a dendritic agglomeration. A large number of cavities is present giving rise to a density that is less than half that of the bulk. A denser film with a few cavities is produced with 1 eV atom^{-1} impact energy: intermixing of clusters and a density of 80% of the bulk is observed. This simulation has been performed cluster by cluster, without considering any deposition rate.

The above results stimulate, although in a qualitative way, several considerations useful for the designing and interpreting a cluster deposition experiment. In particular, the kinetic energy of the impact on the surface is found to strongly influence the structure of the resulting film. The degree of fragmentation, adhesion, coalescence depends on the balance between kinetic and cohesive energy of the aggregates. Moreover, the nature of the surface is also important in determining cluster mobility, creation of defects and, in general, dissipation and redistribution of the impact energy.

A microscopic description of cluster–surface interaction and cluster rearrangement on the surface is far beyond the actual simulation ability and the performance of local experimental probes. As a rule of thumb one can describe the parameters involved by dividing the kinetic energy per atom E/N by the cohesive energy per atom ε of the cluster without considering the cluster–surface interaction [5.14]. This ratio should roughly determine if and to what extent the cluster survives the impact. For $E/N \ll \varepsilon$ the clusters should survive the impact, cluster-cluster coalescence being negligible, and a nanocrystalline material should be formed with a granularity reminiscent of the precursor clusters. On the other hand, for $E/N \gg \varepsilon$ massive fragmentation and coalescence is expected. Increasing further the impact energy, surface modifications induced by cluster impact become important and phenomena such as sputtering and implantation are predominant. In the high energy regimes cluster beams can be useful tools for surface processing and modification, as an alternative to ion beams.

5.2 Diffusion and Coalescence of Clusters on Surfaces

The characterization of chemical and physical adsorption, diffusion, coalescence and disruption of clusters deposited on solid surfaces is important since these processes affect the formation and the evolution of precursors during the first steps in the growth of nanostructured materials.

Cluster beam methods are particularly well suited for studying these aspects since they allow a good characterization of the clusters before the impact, moreover they can be combined to ultra high vacuum and surface characterization equipments.

In an ideal experiment one would require a beam of clusters monochromatic in size, with a well determined initial state in terms of structure, energy and momentum. The surface should be as well characterized in terms of morphology and structure, atomically clean and without defects, or with a well defined defect structure. During the interaction, in such an ideal experiment, one should be able to follow the state of both the clusters and the surface by detecting both the particles on the surface and the ones leaving it.

Such an experiment is practically impossible, but the evolution of cluster beam methods on one side and the recent great achievements in surface physics, including local probes with atomic resolution, are producing a rapid evolution of this field, although a general picture is still lacking.

In this section we will confine the discussion to a few of the most relevant experiments that more directly relate to the study of the early stages of the growth process. Other aspects such as surface modifications induced by cluster impact are discussed in a following section while the growing field of scattering experiments is not considered here since its impact on the synthesis of cluster assembled and nanostructured materials can be considered, up to now, marginal.

Direct observations on the dynamics of clusters on a surface challenge available imaging techniques with local probes. Field ion microscopy (FIM) and scanning tunneling microscopy (STM) and other local probes, with their ability to image a surface with atomic resolution, have been successfully used to monitor morphology and dynamics of clusters on solid surfaces. Even though they often represent the techniques of choice because they produce precious direct information, one should keep in mind their restrictions and limits. For example FIM cannot be adopted for large clusters because the imaged plane on the field ion tip has a diameter typically ranging between 25 to 100 Å [5.15] so that it becomes very difficult to study clusters larger then a few tens of Å. As far as other scanning local probes are concerned, the typical time to acquire an image, that is the time needed for the needle to scan the surface area under observation, is typically too long to observe directly rapid diffusion processes.

The interference between the tip of local probes with the clusters should be taken carefully into account too. It can disturb the aggregates and their dynamics on the surface, affecting the final picture to a degree that may arrive to the point of non-observation of the larger clusters arising from diffusion and aggregation [5.16]. An atomic force microscope, operated in tapping mode [5.17], interferes very little since the effects of displacements of the clusters due to the tip are reduced to a minimum but its resolution is limited (typically 5 nm). In all cases the direct study of diffusion, aggregation and dissociation of the clusters, which is of great interest to understand the processes involved in cluster-assembled materials, is not at all an easy task.

A series of experiments have been performed to elucidate the diffusion and coalescence dynamics of large antimony and gold clusters deposited at low energies on graphite [5.18–5.20]. A gas-aggregation source of the Sattler type is used to generate the antimony cluster beam characterized by two size distributions centered around 2300 atoms (about 5 nm diameter) and 250 atoms (2 nm diameter) respectively. The gold clusters are produced by a laser vaporization source forming a beam with a distribution of sizes centered around 250 atoms.

The experiments are carried out on highly oriented pyrolytic graphite (HOPG) that is an ideal substrate for these studies since it can easily be cleaned by annealing in vacuum (at 500°C for a few hours) a freshly cleaved sample. The surface is atomically flat over large regions (larger than 2000 nm) separated by steps. It is then possible, as we will see in more detail below, to study not only the process on the flat surface but also at the steps which are particularly active sites for clustering.

The morphology of the deposits is characterized as a function of the beam flux, the substrate temperature and the thickness of the deposits by transmission electron microscopy (TEM), scanning electron microscopy (SEM) and atomic force microscopy in tapping mode (TMAFM).

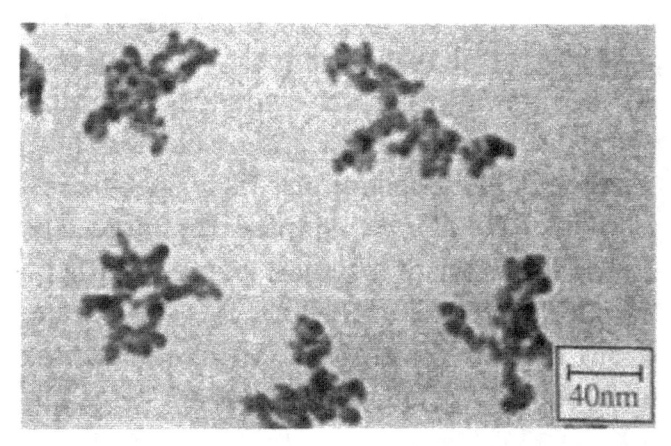

Fig. 5.9. TEM micrograph of antimony aggregates grown on a HOPG surface with a beam of large cluster (2300 atoms). The particles forming the ramified islands have a size similar to that of the incoming clusters. (From [5.19])

Ramified islands, with a fractal shape, are observed. Figure 5.9 shows the features observed by a TEM micrograph on this system. The constituent elements (small round shaped features in the figure) are the deposited clusters since the size of the particles forming the islands matches the dimension of the clusters in the beam. This is a strong indication that the clusters do not fragment nor coalesce during the deposition process. These conclusions are coherent with what one would expect considering that the kinetic energy per atom in the cluster is about 5 meV, well below the cohesion energy of the cluster [5.21].

TMAFM micrographs show (Fig. 5.10) very similar features confirming the round shape of the particles forming the island. The roundness of the deposited clusters can be understood if their interaction with the graphite is so weak that it would not affect their morphology. This is confirmed by molecular dynamics calculations [5.22].

The growth process depends on the coverage at fixed beam flux. Figure 5.11 shows the evolution of the island density. Nucleation processes, where two clusters moving on the surface collide and form an island, are the most probable at very low coverages, giving rise to a fast increase of the number of islands in the first part of the graph in the figure. When the density of islands becomes large enough, the incoming cluster migrating on the surface will have an increasing probability of being captured by an already formed island, so that there is competition between the growth of the island and the increase in density of islands.

This gives rise to the region characterized by a slower increase of the density of islands at surface coverages between about 6% and 10%. At coverages above 10%, most of the incoming clusters are captured by already formed islands, giving rise to a pure island growth regime without the formation

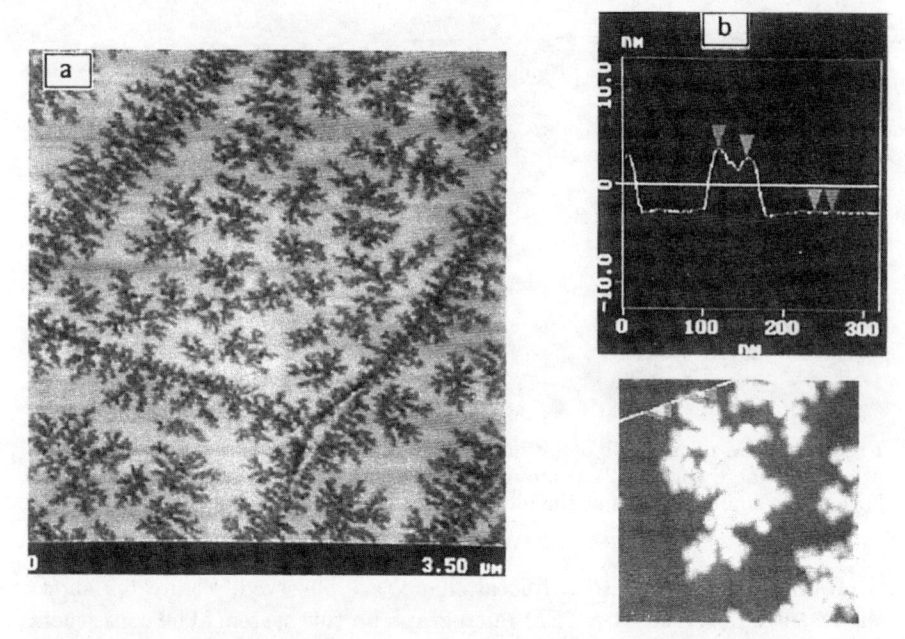

Fig. 5.10. (a) TMAFM image of a film of antimony grown from large clusters; (b) cross section of the clusters forming the island while at the lower right is shown a magnification of an island where the size of the clusters component is visible. (From [5.19])

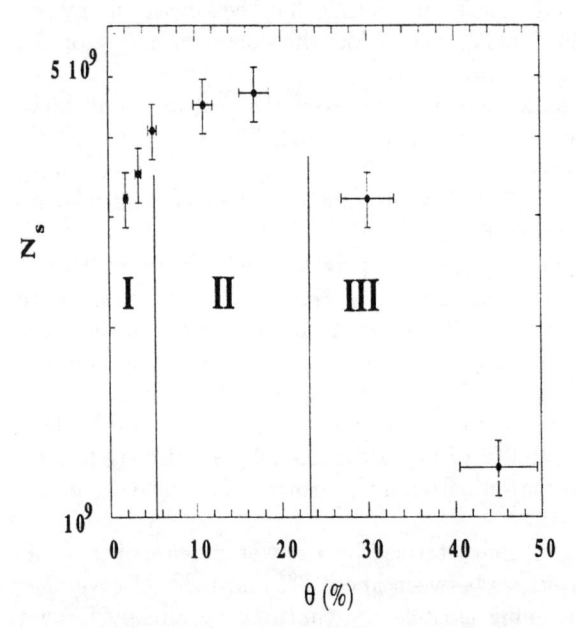

Fig. 5.11. Island density (in number of islands per cm^2) as a function of the surface coverage. (From [5.19])

of newer ones, so that their density becomes stationary. Finally, when the island–island distance becomes comparable to the linear dimensions of the islands (coverages of about 25–30%), the onset of coalescence between islands induces a decrease of the islands density. The evolution of the morphology and density of islands as a function of coverage is consistent with the hypothesis that the sticking of a cluster to an island is irreversible, even though an intra-island mobility cannot be excluded.

When the beam flux is increased, the major observed effect is a reduction of the island density and a corresponding increase in their ramification. This behavior is consistent with a picture where the aggregation in islands is simultaneous with the deposition and to the diffusion of the clusters. Moreover, the presence of a significant density of active sites (due, for example, to defects and contamination), which act as nucleation centers on the surface, can be excluded.

The dependence of these features on the size of the incoming clusters has been investigated in the experiments with small cluster sizes (distributions in the beam centered at 250 atoms/cluster). They show a clear evidence of coalescence between clusters. This process coexists with island formation, the overall shape of which does not change significantly from the case of larger clusters. Coalescence is demonstrated by the observation that the particles forming the islands have a size (3–5 nm) that is from 5 to 10 times larger than that of the incoming clusters. Gold clusters of similar size give rise to similar features showing that coalescence between small clusters is not specific to antimony. These major trends are confirmed by other experiments on Ag_n ($n \approx 300$) clusters deposited on graphite [5.23].

The overall picture that emerges is that a growth mechanism based on a sort of paving of the surface is not adequate. A more realistic growth mechanism should instead be based on the evidence that clusters diffuse on the surface while they do not always merge when they touch each other. From this point of view, another very interesting feature that is clear from the experiments is the existence of a sort of a "critical size for coalescence". Clusters having sizes larger than the critical one do not undergo a coalescence process at their encounter. On the contrary, if even only one of the colliding clusters has a size smaller than the critical one they will merge forming a larger particle.

It is interesting to note that the critical size corresponds roughly to the size of the supported clusters forming the islands [5.24]. Even though a complete understanding of this feature is not available, an indication of a possible interpretation has been proposed. The existence of such a critical size could be related to a liquid–solid transition of the cluster that, as is well known depends strongly on its size. Liquid-like clusters can easily coalesce and in so doing will grow up to the critical size for becoming solid. Further collisions will not cause any coalescence because solid clusters have a much lower propensity to merge. Unfortunately the available data do not give enough information to definitely confirm this interpretation.

These experimental observations have been compared with a detailed model including deposition, diffusion and cluster aggregation (DDA models), which could answer basic questions concerning the diffusion dynamics of the clusters, the mobility of the islands, the microscopic diffusion mechanisms and which are the factors governing the coalescence of clusters [5.25–5.29].

The DDA model proposed by Jensen et al [5.26] assumes the following mechanisms for the three relevant processes. The deposition occurs with a flux F at random positions on the surface; at every diffusion time τ all the particles are chosen randomly and move by one diameter in random directions; the aggregation in islands occurs when two clusters occupy neighboring sites. Further assumptions, more or less implicit in the model, regard the sticking coefficient that is assumed to be 1 between both two clusters and each cluster and the surface. Furthermore, there is assumed to be no cluster dissociation from the island nor diffusion along the island border while no coalescence is assumed to occur between clusters. The diffusive motion is Brownian. These hypotheses are supported, or indirectly confirmed, by the mentioned experimental observations and the temperature dependence of the morphology of the islands.

The model defines the normalized flux Φ as the number of clusters deposited per site per diffusion time:

$$\Phi = \frac{F\tau\pi d^2}{4}, \tag{5.2}$$

where F is the flux per unit time per unit area, and d is the average diameter of the incident clusters. The cluster diffusion coefficient D is given by

$$D = \frac{d^2}{4\tau}, \tag{5.3}$$

and the maximum density of islands (saturation density) N_{isl} in the model becomes a function of the normalized flux:

$$N_{\text{isl}} = \frac{\Phi^\chi}{2}. \tag{5.4}$$

From the comparison with experiments the parameter χ is determined to be 0.36 [5.19]. From (5.2), (5.3) and (5.4), the expression for the diffusion coefficient assumes the form:

$$D = \frac{F\pi d^4}{16}\left[2N_{\text{isl}}\right]^{1/\chi}. \tag{5.5}$$

From (5.5) and the experimental dependence of N_{isl} as a function of the temperature, the cluster diffusion coefficient can be studied.

As shown in Fig. 5.12 the diffusion is very rapid on the graphite surface in spite of the large size of clusters. The reason for this should be sought in the interaction with the substrate. The mobility on other surfaces is in fact much smaller as has been demonstrated by similar experiments dealing with

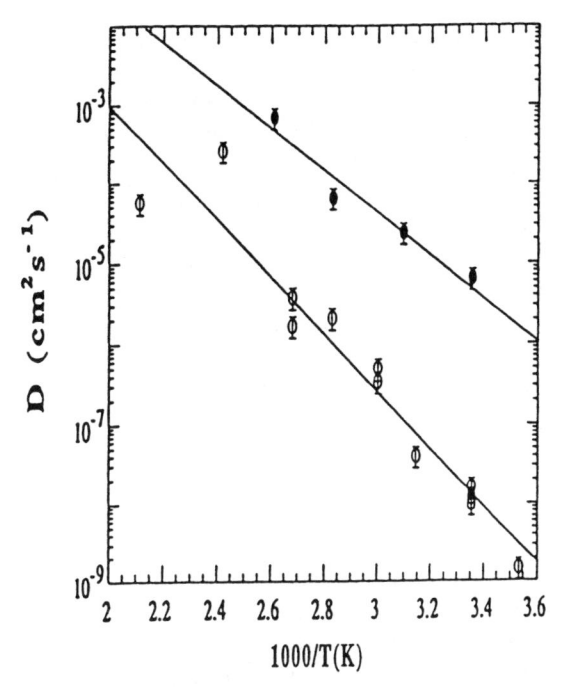

Fig. 5.12. Temperature dependence of the diffusion coefficient of Sb_{2300} (o) and Au_{250} (•) clusters on the surface of HOPG. The full lines represent the corresponding best fitted Arrhenius plots: for Sb_{2300}, $E_a = 0.7\,\text{eV}$ and $D_0 = 10^4\,\text{cm}^2\text{s}^{-1}$ while for Au $E_a = 0.5\,\text{eV}$ and $D_0 = 10^3\,\text{cm}^2\text{s}^{-1}$. (From [5.20])

gold clusters on KCl and with silver clusters on silver [5.30, 5.31]. For these systems the diffusion coefficients are up to 6 orders of magnitude smaller.

The results of the analysis of these experiments can be summarized in the following way. Large antimony clusters (2300 atoms) are very mobile on the graphite surface and, when they collide, do not coalesce to form a larger aggregate. In the case of the smaller clusters (250 atoms), both gold and antimony, coalescence becomes probable. Cluster diffusion appears to be a thermally activated process with large values of D_0 (up to $10^4\,\text{cm}^2\text{s}^{-1}$). The island formed by the collision of two clusters is not mobile even if it is formed by two small clusters and its size is about 500 atoms, which is much smaller than the large antimony clusters (2300 atoms) that, instead, appears to be very mobile.

These features strongly support the picture of a cluster moving as a quasi-rigid structure and seriously question the validity of a mechanism of cluster diffusion as resulting from the combined sequence of "atomic" movements of the cluster constituents as proposed by the "periphery diffusion" and "evaporation–condensation" models [5.30]. The fractal shape of the islands can be explained by the combined effect of irreversible sticking of the incoming clusters on the islands and the very low probability of coalescence.

The dynamics of silver clusters produced by a gas-aggregation source, ionized by a hot-cathode plasma, size selected (ranging from 50 to 250 atoms) by a Wien filter and deposited on a surface of HOPG is reported in [5.32]. As in the experiments examined before, the results show clearly the formation of ramified islands but with the remarkable feature that they are formed by particles with a narrow size distribution centered around 14 nm that the authors define as "universal" diameter. This is a feature that does not depend on the size of the incoming clusters and it implies a coalescence process of at least 100–500 of them (the estimated size is in excess of 20 000 atoms). The mechanism behind this behavior should depend on the strain that particles larger then 14 nm will suffer because of the large lattice mismatch with the surface structure. The formation of strained islands have been predicted by molecular dynamics calculations on Ag clusters absorbed on the Pt(111) surface, where the interatomic potential constrains the Ag atoms to sit in registry with the platinum lattice [5.33].

The very important role played by defects and, in particular, by atomic steps in coalescence process on surfaces is also investigated [5.34]. To this end a Knudsen source is used producing small or no clusters at all to form nanostructures induced by the surface diffusion at the steps at submonolayer regimes. A sequence of SEM micrograph shows the coalescence at the atomic defects present on a HOPG surface. For low deposition rates, clusters grow only at the atomic steps with a narrow size distribution centered around 10 nm. Figure 5.13 shows an SEM micrograph where steps of the graphite surface decorated by the formed clusters are clearly visible. Clusters show

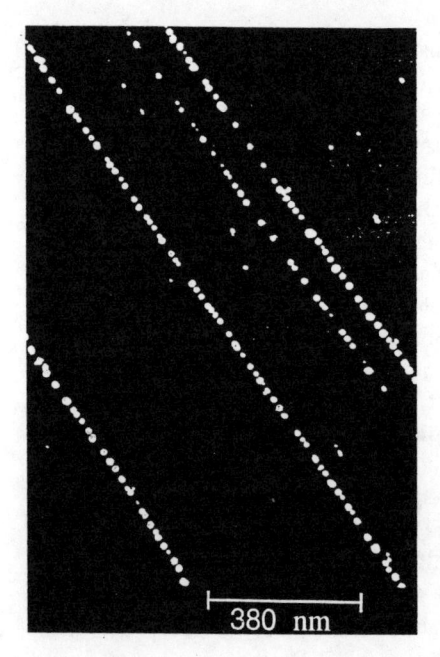

Fig. 5.13. SEM micrograph of Ag clusters formed on HOPG. The surface has been exposed to Ag vapors for 2 s at a rate of 2×10^{13} cm^{-2}s^{-1} at a surface temperature of 118 °C. (From [5.34].) The missing elements in the second chain from the top are probably the ones appearing over the terraces that have diffused away from the step where the chain is formed

the tendency to align along the surface steps in a quasi-unidimensional chain. The diffusion processes of the clusters depend on the type of defects acting as strong localization centers of the growing clusters.

Kern and co-workers [5.35] have studied the energy dependence of the interaction of silver clusters with the Pt(111) surface and the feasibility of deposition of size selected clusters to synthesize nanostructured materials. The preparation and selection of a beam of clusters of a specific size (7 and 19 atoms in the specific case) is achieved by a sputtering source and a quadrupole mass selection filter [5.36]. Ultraclean conditions and an STM *in situ* guarantee results free from spurious effects due to contamination. Since the kinetic energy range achievable does not allow to work in a regime free from surface damage and cluster disruption, experiments are also carried out by covering the surface by a layer of Ar that acts as a buffer that allows a "soft" slow down of the clusters.

In order to characterize the specific features of cluster deposition, the authors compare these results to the thermal growth of islands arising from the deposition of atoms of Ag deposited on the surface.

The nanoaggregates formed by deposition of Ag_7 clusters look very different: the atoms give rise to strongly ramified structures while the clusters produce very compact islands.

The deposition on the clean surface is carried out at low temperatures (80–90 K) where the island structures are stable and localized. An interesting aspect is that the imaged islands are always 2D (one atom thickness) even when the precursors are 3D clusters. This behavior is due to the strong interaction with the Pt surface where the absorption energy is sufficient to activate the transformation of morphology. When the temperature of the surface is increased over room temperature, the nanostructures formed are annealed and the deposited silver condenses at the step edges of the Pt surface. The annealing of the structures grown by the evaporated atomic Ag is complete (Ostwald ripening [5.37]) while some of the nanostructures produced by cluster deposition remain. This is due to the combined effect of surface damaging and fragmentation induced by the kinetic energy of the cluster.

In fact, the direct landing on the clean surface of Ag_7 clusters with 20 eV energy (i.e. 2.9 eV per atom) produce surface defects that act as pinning centers for the diffusing atoms formed by the partial fragmentation of some of the impinging clusters. Such defects stabilize the nanoaggregates that would be stable and not mobile on the surface even at room temperature so that they do not anneal.

Depositions of Ag_7 clusters at the same beam kinetic energy but through an adsorbed Ar buffer (about 10 monolayers thick) will give rise to nanoaggregates with the same morphology. The experiment is carried out by adsorbing the Ar buffer layer before the exposure to the cluster beam at about 25 K. After the deposition of the clusters the surface is annealed at 90 K temperature where the Ar buffer layer evaporates and the resulting surface can be directly imaged by the *in situ* STM. The interesting result is that, when brought to

300 K, the silver aggregates anneal completely without leaving any traces of nanoaggregates. This leads to the conclusion that no defects or fragmentation of clusters is produced: soft landing of clusters, via energy dissipation into a rare gas buffer layer, seems to be a viable technique to deposit clusters on surfaces.

Another interesting conclusion is that the growth of nanostructures can be stabilized by controlling the density of defects on the surface. The kinetic energy of $1\,\mathrm{eV\,atom^{-1}}$ is established as the limit over which the damaging and dissociation effects become sizable.

The state of the art of cluster beams–surface interaction shows a field that is still in its infancy but already demonstrates how promising it is in terms of processes and regimes. Cluster beams will play a crucial role in gaining a better understanding of the related phenomena and this is a critical step to control the early stages of the growth process. Major progress in this direction will produce important steps forward in the direction of producing nanostructured films with tailored properties.

5.3 Low-Energy Deposition

Nanostructured materials can be described as consisting of two structural components: a crystalline one formed by small single crystals on a scale ranging from a few to roughly one hundred nanometers, and an intercrystalline component which is formed by boundaries between the crystallites. This component is characterized by a reduced atomic density and interatomic spacing, deviating from those of a perfect crystal lattice [5.38, 5.39].

From this point of view nanostructured materials lie in a region between crystalline and amorphous solids showing properties common to both states. Moreover, they also display unique features which are related to confinement and interface effects. Confinement effects are due to the fact that electronic and vibrational excitations have characteristic lengths comparable to the diameter of the crystallites. In this case substantial modifications of the phonon density of states and of the electronic structure should be expected (see below). Interface effects, due to the large number of grain boundaries, affect the mechanical properties of the material; the high porosity can also have a strong influence on the catalytic properties.

Among different production methods [5.38], cluster beam deposition is particularly suitable for the synthesis of thin films, layered and composite structures. With this technique it should be possible to control the degree of crystallinity of the film without any thermal treatment, the size and the density of the embedded particles in case of composite structures.

The growth of films via low-energy cluster beam deposition can be viewed as a random stacking of particles as for ballistic deposition [5.40]: the resulting material is characterized by a low density compared to that of films assembled atom by atom and it shows different degrees of order depending on the scale of observation. The characteristic length scales are determined by cluster dimensions and by their fate after deposition: clusters in beams are characterized by a finite mass distribution and by the presence of several isomers with different stabilities and reactivities. Once on the substrate stable clusters can survive while reactive isomers can coalesce to form a more disordered phase. The physico-chemical properties of nanostructured films depends on the coexistence of different characteristic lengths and on their interplay: this constitutes the peculiarity of these systems.

The characterization of nanostructured materials is performed with techniques that also have typical length scales. This makes the interpretation of the results more delicate and should encourage a multi-technique approach to the study of cluster-assembled materials (see Chap. 6). Concerning the effects of localization, neutron scattering or resonant inelastic nuclear gamma-ray scattering are very effective to characterize the phonon density of state of a material, unfortunately they require bulk quantities of the sample often difficult to obtain with cluster beams [5.41]. Electronic spectroscopies can be elemental sensitive on a very short scale (atomic), on the other hand their lateral resolution is not sufficient to separate the contribution of individual clusters.

Local probe microscopies have already been discussed for the characterization of low-coverage cluster layers; for cluster-assembled materials in the ballistic regime, the roughness of the films and their fragility make the use of these probes quite difficult. In particular, AFM should be used in non-contact modes in order to avoid modifying the sample [5.42].

Raman and Brillouin spectroscopy can provide information about the confinement of phonons on a scale range up to hundreds of nanometers. These techniques can be very useful for the characterization of nanostructured materials, as we will discuss in the following, however, they should always be coupled with other diagnostic techniques to extract unambigous information on confinement and particle size distribution.

5.3.1 Cluster Networks and Porous Films

Semiconductor Clusters. Semiconductor-based nanocrystalline materials are the subject of considerable interest driven by their optical and electronic properties [5.43]. Group IV-based structures, for example, consisting of Ge and Si nanocrystals embedded in matrices or free-standing, exhibit visible photoluminescence that is sensitively dependent on the particle size and size distributions [5.44–5.47]. Despite the efforts devoted to the characterization of these systems and to their electronic structures, many aspects remain to be clarified. The confinement of elementary excitations in semiconductor nanocrystals has a strong influence on the photoluminescence behavior and more generally on their linear and non linear optical properties [5.48]. For an unambiguous characterization of the role played by structure and dimensions in determining the quantum confinement effects, it is necessary to combine the preparation of well characterized nanostructures with the ability to follow their evolution along the different steps of the synthetic routes.

A wide variety of techniques have been used to produce semiconductor nanocrystals including: gas evaporation [5.49, 5.50], rf magnetron sputtering [5.44, 5.51–5.54], chemical vapor deposition [5.55], chemical reduction of metastable species [5.56–5.59], oxide reduction in zeolite [5.60], inorganic solution-phase synthetic routes [5.61], ion implantation and annealing [5.62]. Molecular and cluster beam deposition have also been proposed [5.63–5.65].

Among these different synthetic routes, CBD may be very interesting for the production of samples characterized by single-sized particles or assemblies of particles with a well defined size distribution. Moreover, it may also allow control over the crystalline state of the particles and on their interaction with the embedding matrices.

Ge particles are prepared by ionized cluster beam deposition, however, no attempt has been made to correlate the cluster beam characteristics with the deposited particles. The lack of any reliable beam characterization in terms of cluster mass distribution (see below) makes it very difficult to draw general conclusions on the effectiveness of this technique.

The situation concerning silicon is different. Since CBD seems to allow a good control over particle sizes and shapes it may offer some advantages in the preparation of nanocrystalline samples over more conventional techniques for the preparation of porous silicon which are based on electrochemical etching or plasma enhanced chemical vapor deposition from an rf discharge of silane [5.43].

An intense source of silicon particles has been realized by Ebrecht and co-workers. Silicon cluster molecular beams are produced by using a gas-flow reactor where SiH_4 is decomposed by continuous or pulsed CO_2 laser irradiation [5.66–5.68]. A schematic view of the beam apparatus and typical mass spectra are shown in Fig. 5.14 and Fig. 5.15, respectively.

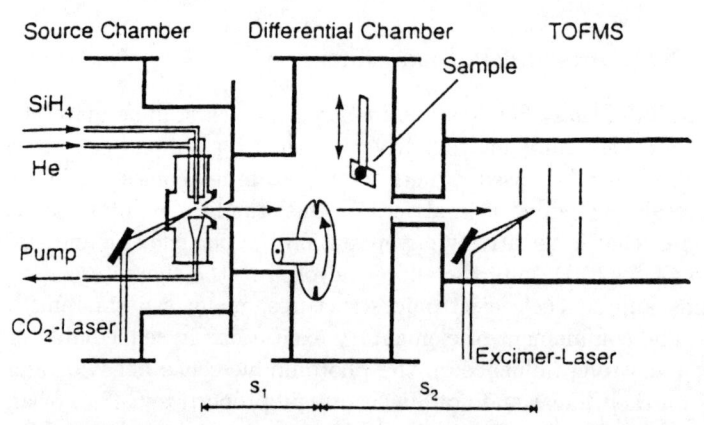

Fig. 5.14. Schematic side-view of the molecular beam apparatus. (From [5.66])

5.3 Low-Energy Deposition   145

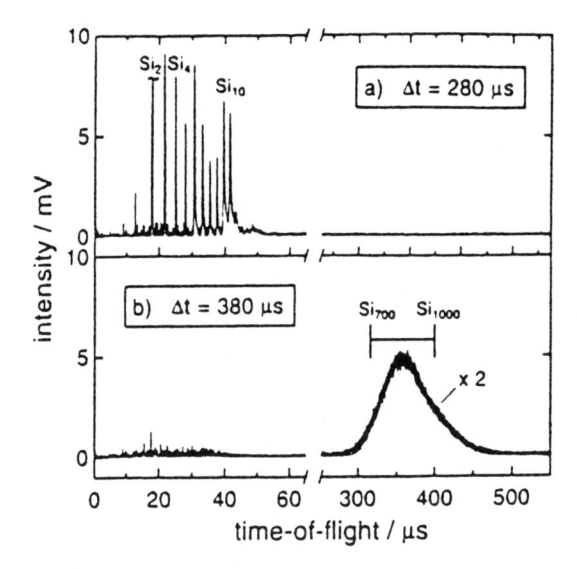

Fig. 5.15. TOF mass spectra of Si clusters taken at different delays between the reaction-driving laser and the ionizing laser. (From [5.66])

As discussed in Chap. 3, the mass distribution of the produced clusters depends on the residence time in the reaction zone (Fig. 5.15). A mechanical velocity selector can thus select a specific mean size of cluster distribution, allowing the synthesis of nanocrystalline films characterized by different precursor dimensions.

Raman spectroscopy has been used to obtain the structural characterization of the nanocrystalline films. The Raman spectrum of crystalline silicon is characterized by a Γ'_{25} optical phonon at 521 cm^{-1} with a natural width of ≈ 3.5 cm^{-1} [5.69,5.70]. In amorphous silicon the $q \simeq 0$ selection rule does not apply and the Raman spectrum resembles the phonon density of states with a broad peak centered around 480 cm^{-1}. Samples characterized by incomplete or incipient crystallization, show a Raman spectrum which is somehow intermediate between the two extreme cases: a broadening and red-shift of the 521 cm^{-1} peak is observed together with the appearance of a broad feature around 480 cm^{-1} [5.70]. Microcrystalline silicon films, characterized by a grain size of the order of a few tens of nanometers, show a similar behavior with the lineshape and red-shift of the 521 cm^{-1} line depending on the crystallite size.

Figure 5.16 shows the size distributions and Raman spectra of nanocrystalline films produced with CBD, compared with a c-Si(111) spectrum [5.68].

These results are in good agreement with theoretical models of small Si particles [5.69], the mean cluster size, extracted from fitting of Raman spectra, indicates that the size distribution in the beam is not substantially changed after deposition.

By analogy with porous silicon, the nanocrystalline films luminesce with a luminescence peak shifting to higher energies as the particle diameter de-

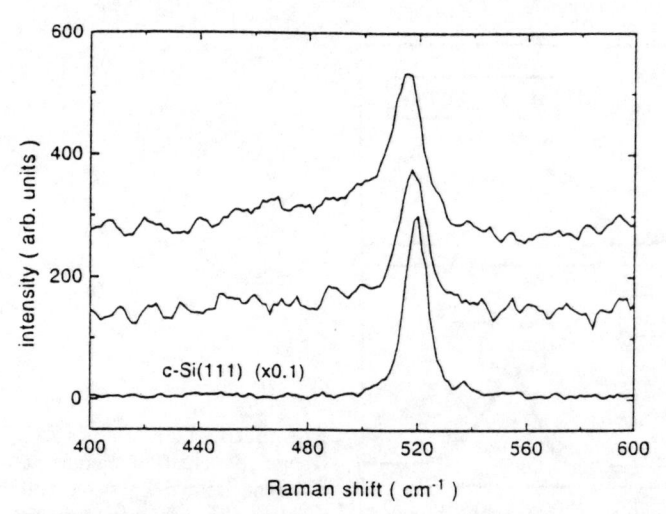

Fig. 5.16. Raman spectra of cluster-assembled silicon films with different size distribution. The size distribution for the top spectrum is peaked at 4.63 nm and for the middle one at 7.03 nm. The bottom spectrum is that of c-Si(111). (From [5.68])

creases. The agreement is only qualitative since the cluster deposited films luminesce at considerably shorter wavelength. This behavior could be explained by considering the shape of the particles (round-shaped compared to rod-like structures of porous silicon).

The absence of coalescence is in agreement with similar Raman and photoluminescence results on Si particles embedded in a SiO_2 matrix [5.47]. For very small clusters ($d < 2$ nm) the quantum confinement model cannot be applied and the interpretation of Raman and PL spectra becomes difficult without an independent determination of the cluster size distribution.

Silicon thin films have also been deposited using a cluster size distribution peaked at very small masses (< 100 monomers per cluster) [5.71, 5.72] (Fig. 5.17).

The Raman spectra are comparable with those of a disordered phase that can be accounted for by the presence of very small particles, however, the luminescence spectra are typical of particles with diameter between 3 and 5 nm [5.68]. Some of the observed features can be attributed to the presence of particles with a structure different from that of bulk silicon, however, experiments using different cluster distributions in the beam should be performed to confirm this hypothesis [5.72].

Carbon Clusters. The assembling of nanocrystalline carbon thin films by cluster beam deposition has been investigated both experimentally and theoretically [5.24, 5.73–5.75]. The control of beam parameters such as the kinetic energy, is expected to give the possibility of tuning the sp^2/sp^3 ratio of the film by analogy with other deposition techniques. Moreover, the control on the cluster size distribution should allow the introduction of nanoscopic

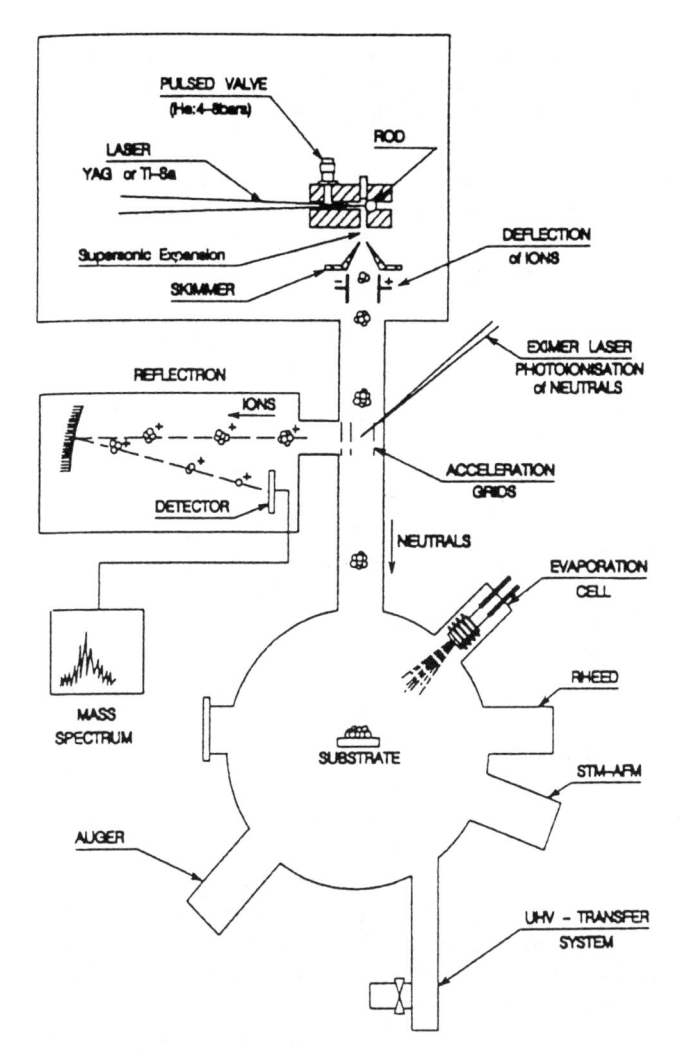

Fig. 5.17. Block diagram of the cluster beam deposition apparatus used for silicon cluster deposition. (From [5.72])

structures (chains, fullerene-like units, etc.) otherwise difficult to create by assembling the film atom by atom. Novel structural and functional properties should be displayed by film retaining some of the cluster characteristics [5.76, 5.77].

Nanocrystalline carbon thin film can be produced by depositing cluster beams generated by plasma cluster sources. Melinon and co-workers [5.20, 5.24] have characterized the structural and mechanical properties of films obtained with a LVCS at different operating conditions which should correspond to different cluster mass distributions. The observation of mass spectra peaked at C_{20}, C_{60} and C_{900} are interpreted as an indication of different size distributions, however, a systematic characterization of the cluster beams is not reported by taking into account the residence time of the clusters in the source.

Low-density films ($\sim 1\,\mathrm{g\,cm^{-3}}$), characterized by a granular structure on a scale of several tens of nanometers, show Raman spectra evolving from a graphitic character when large clusters (C_{900}) are used, to an amorphous character as the cluster size distribution is reduced to a few tens of atoms per cluster [5.24].

Raman spectroscopy can be used to characterize the sp^2/sp^3 presence in carbonaceous materials on a nanometer scale, however the ratio between three-fold and four-fold coordinated carbon atoms cannot be inferred solely from Raman spectra, unless one uses ultraviolet excitation sources [5.78]. Melinon and co-workers have assumed, from Raman spectra taken with excitation at 532 nm, that films formed with small clusters have an sp^3 character due to the formation of diamond-like phase by the assembling of cage-like C_{20} clusters. Unfortunately, the available experimental evidence does not support this hypothesis. Despite the uncertainty on the final film structures, these studies showed the existence of a "memory effect" that is the dependence of the final structure of the film from the mass distribution of the precursor clusters.

Nanostructured carbon films can also be deposited using cluster beams produced by a pulsed discharge source (see Chap. 3) [5.73, 5.74].

The characterization of cluster mass distributions and charge states in the beam, as a function of the source parameters, has been performed with a linear time-of-flight mass spectrometer. Neutral clusters are ionized with the fourth harmonic (266 nm) of a Nd:YAG laser. Cluster fluxes are characterized with a quartz microbalance and with a Faraday cup. In Fig. 5.18 a typical mass spectrum of neutral carbon clusters is shown: the mass distribution is peaked around 800 atoms per cluster, with contributions from aggregates up to about 2500 atoms per cluster.

The center of mass of the size distribution and charge state are strongly influenced by the presence of the thermalization cavity. The clusters residence time depends on parameters such as the pressure reached in the cavity, its volume, and the conductance of the nozzle. The spectrum in Fig. 5.18 has been recorded with a delay time T_d between the arc discharge and the laser ionization of 1100 μs. This delay is the convolution of the residence time of the cluster in the source with the time necessary for the cluster to reach the TOF mass spectrometer. Mass spectra can be recorded for a very wide range of T_d (several milliseconds), showing that the center of mass of the size distribution is slightly influenced by the residence time. For very large T_d, although clusters of several thousands of carbon atoms contribute to the mass spectra, no substantial changes in the position of the center of mass of size distribution are observed. The spectrum in Fig. 5.18 takes into account all the spectra produced up to $T_d = 15\,\mathrm{ms}$ after the discharge.

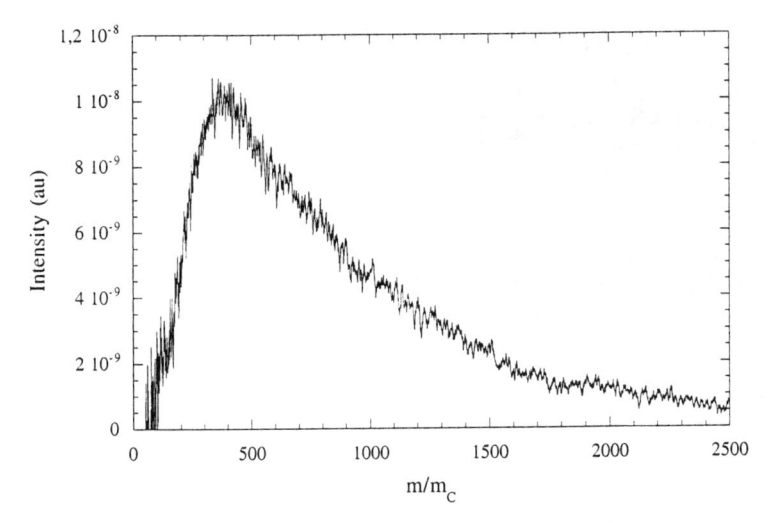

Fig. 5.18. Typical mass distribution of carbon clusters produced by a pulsed arc source. (From [5.73])

Using a deposition rate of several nm min^{-1} and kinetic energies of a fraction of eV per atom, porous films with a density and a porosity similar to those reported in [5.24] are produced (Figs. 5.19 and 5.20).

Visible and UV Raman spectra show that films deposited with large clusters are characterized by the presence of a high number of distorted sp^2 bonds

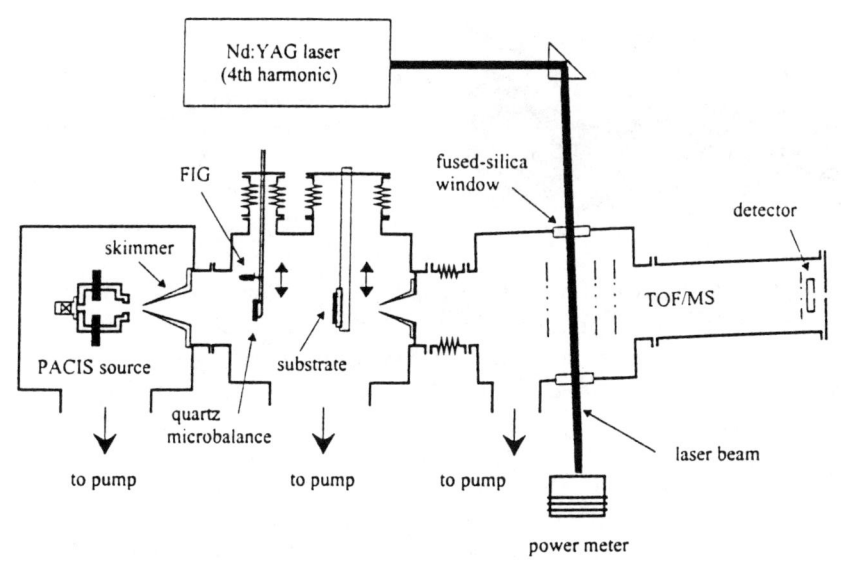

Fig. 5.19. Schematic view of the apparatus for thin film deposition. (From [5.73])

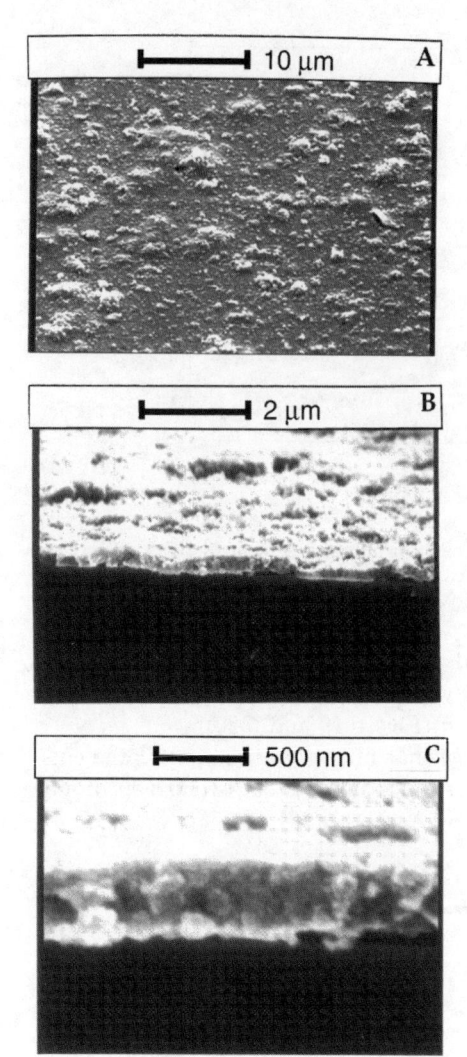

Fig. 5.20. SEM micrographs of nanocrystalline carbon films. (**A**) low-magnification micrograph of the surface. (**B**) and (**C**) high magnification micrographs of a section of the film. (From [5.73])

(Fig. 5.21), whereas when depositing small clusters (few tens of atoms) sp^3 and sp coordination is present. These results are in qualitative agreement with XPS measurements, although photoemission spectroscopy is sensitive to a scale smaller than that of Raman. For large clusters, optical spectra confirm the disordered sp^2 structure and show the presence of a gap of 0.6 eV.

Although Raman spectroscopy is a powerful tool for studying the structure of carbon-based materials, the information is restricted to the scale of a few nanometers. Experimental characterization and theoretical modeling of cluster-assembled granular materials have to face the problem of cluster coalescence and of their organization in structures spanning length scales

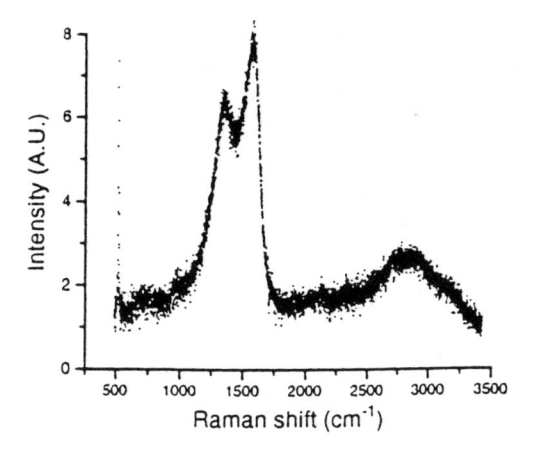

Fig. 5.21. Raman spectrum of a film deposited at room temperature on a silicon substrate. (From [5.73])

from the nanometer up to the micrometer. The different structures in which the precursor clusters are organized needs experimental probes sensitive to the different length scales typical of intra-cluster and inter-cluster interactions. In order to study the organization of clusters on a scale of hundreds of nanometers, which is the typical scale of thermally excited long wavelength acoustic phonons, Brillouin light scattering can be used [5.80]. Films of graphite [5.81], polycrystalline diamond [5.82], diamond-like a-C:H [5.83], C_{60} (fullerite) [5.84], and phototransformed C_{60} [5.85] have also been studied by Brillouin scattering. Usually this spectroscopic technique is applied to homogeneous compact films with perfect (atomically flat) surfaces and buried interfaces. Particularly challenging is the attempt at obtaining and interpreting Brillouin spectra of films with a rough surface and/or a granular or porous structure. The extraction from the spectroscopic data of the elastic properties of these systems can be based only partially on what is known in the case of good films and a complete theory is still lacking. From Brillouin spectra the elastic properties of the material can be extracted with a higher accuracy compared to nanoindentation measurements which require a complex analysis and a careful interpretation [5.86].

Thick films (thickness $\geq 0.8\,\mu m$) and thin films (thickness $\simeq 0.1\,\mu m$) have been examined [5.80,5.87]. In thick films only damped bulk acoustic phonons with a typical wavelength λ_{ph} of the order of 170 nm have been detected, giving rise to coherent Brillouin scattering of laser light [5.88]. This indicates that for a length $d \geq \lambda_{ph}$ the films can be modeled as a continuum with translational invariance and effective elastic constants although structural disorder at smaller scales scatters the phonons significantly. The presence of a rather strong central peak in the spectra may be ascribed to non propagating (overdamped), or to confined vibrational excitations within the films, probably connected with different characteristic correlation lengths less than d. The most damped bulk acoustic phonons may be coupled to the confined modes by a relaxation mechanism.

152   5. Thin Film Deposition and Surface Modification by Cluster Beams

In thick films, a broad peak is clearly visible at about 17 GHz together with a strong central peak about 10 GHz wide. Varying the incidence angle θ does not shift the peak position in frequency: a behavior typical of bulk phonon peaks. The 17 GHz peak has been attributed to the bulk longitudinal acoustic phonon of the film material. This attribution is also supported a posteriori by the numerical value of the elastic moduli of thin films which agree with their graphitic-like structure as determined by Raman spectroscopy [5.87].

In Fig. 5.22, the surface Brillouin spectrum of a thin (100 nm thick) film is shown for an incidence angle of $\theta = 50°$. At least three peaks are clearly visible below the transverse threshold of the silicon surface, superimposed on a central peak which is not wide enough to be the silicon one [5.87].

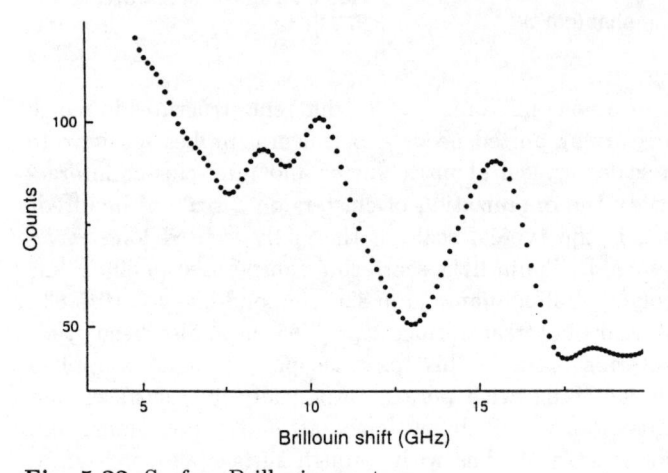

Fig. 5.22. Surface Brillouin spectrum

All these peaks exhibit the characteristic surface frequency variation with the incidence angle although being anomalously broad with respect to normal surface peaks. Fitting the spectral shifts of the resonances in the theoretical spectra to experimental data, the effective elastic constants of the films were obtained. The shear modulus and the bulk modulus turned out to be respectively $\mu \simeq 4.0$ GPa and $B \simeq 3.67$ GPa, corresponding to a Poisson's ratio ν of about 0.10.

The numerical value of μ is in the range of the C_{44} elastic constant of hexagonal crystalline graphite ($C_{44} = 2$–5 GPa) and equals the value of its shear modulus along the C axis as reported in [5.89, 5.90]. This last result can be compared with Raman measurements [5.73, 5.74] which point out the mainly sp^2 carbon bonding present in the disordered granular structure of the films on a nanometer scale. Yet the film material is not elastically identical to nanocrystalline graphite: in fact the values of B (and, consequently, ν) are significantly different [5.89, 5.90].

Metallic Clusters. Memory effects are also observed in metallic nanocrystalline films. For these systems an important factor influencing the properties of the film is cluster mobility on the substrate, as can be seen in the case of antimony and bismuth films deposited using gas-aggregation sources [5.14,5.91]. The structure of Sb films depends on the precursors used: for small Sb_N clusters ($N \leq 300$), the deposited film crystallizes when the percolation threshold is exceeded, while the size of the supported clusters varies with the deposition rate [5.92]. For large Sb_N ($N \geq 1000$), the films are amorphous even when the percolation threshold is exceeded, in contrast to films deposited from molecular precursors [5.14]. The behavior of films grown with large clusters may be explained in terms of low mobility and low coalescence of the aggregates favoring the growth of the film as a filling of a random network in a percolation model. TEM analysis of the film morphology reveals a random stacking of incident clusters which form a low-density network. The critical percolation threshold thickness for electrical conduction is significantly lower for cluster-assembled films compared to traditional ones.

Bi_N clusters show a different behavior with an increase of the percolation thickness compared to standard Bi films. This may be explained by assuming a high mobility of Bi_N clusters, comparable with Bi molecules [5.91].

Beams of covalent clusters have been produced and deposited on various substrates, the resulting films have been characterized in their structural and functional properties. Supported particles observed by TEM show a distribution peaked around 4 and 3 nm for Ni and Co, respectively, indicating coalescence of the incident clusters. XPS and Rutherford backscattering on films protected with an a-C layer, indicate the presence of a significant oxygen contamination of the films. The presence of oxygen can be due to contaminations in the LVS and/or in the deposition chamber (pressure during deposition at 10^{-7} Torr). X-ray grazing incidence scattering reveals the presence of fcc phases for both Ni and Co, while the presence of oxygen in the particles increases the interatomic distances of the outer atomic shells as suggested by extended x-ray absorption fine structure (EXAFS) measurements.

The magnetic properties of these composite films have been characterized, however, the interpretation of the influence of the particle nanostructure is not straightforward: size distribution, departures form ordered structures, surface oxidation, and particle concentration affect the individual and collective magnetic response. Unfortunately, these parameters are not yet controlled to the extent of allowing unequivocal interpretation of the experimental results and comparison between samples deposited with different sources. Squid magnetization measurements on Fe, Ni, and Co clusters show a bulk-like behavior of the magnetization for Fe, a strongly reduced value for Ni and an increasing one for Co [5.93,5.94]. The behavior of thick (100 nm) Co cluster films can be interpreted in the framework of a random anisotropy model where the correlation length of the local anisotropy axes is the mean radius of the grains [5.20].

5.3.2 Composite Nanocrystalline Materials

Cluster beams seem to offer several advantages for the production of composite materials by embedding the aggregates in a matrix. The dispersion of clusters in a codeposited matrix limits the particle coalescence and preserves the original size distribution. Homogeneous dispersion of embedded clusters can be obtained with a size distribution which is independent of particle concentration, in contrast to systems prepared by precipitation [5.20]. A large variety of materials can be chosen for the cluster–matrix complex.

Composite films of cobalt clusters embedded in various matrices (Ag, SiO_x, Cu) have been deposited by a LVCS and a plasma gas-aggregation source [5.20, 5.95]. Co clusters with a mean diameter of 3 nm embedded in Ag and SiO_x show, for concentration below 25%, a superparamagnetic behavior. At larger concentrations the onset of a ferromagnetically ordered state is observed [5.20].

Haberland and co-workers [5.95] have deposited cobalt clusters in a copper matrix at different size distributions and concentrations. Magnetic x-ray circular dichroism has been used to measure the room temperature magnetization of the clusters, showing a dependence on cluster size and concentration. All the samples show, however, a considerably lower magnetization compared to a cobalt thin film up to magnetic fields of 4 T. This behavior may be a sig-

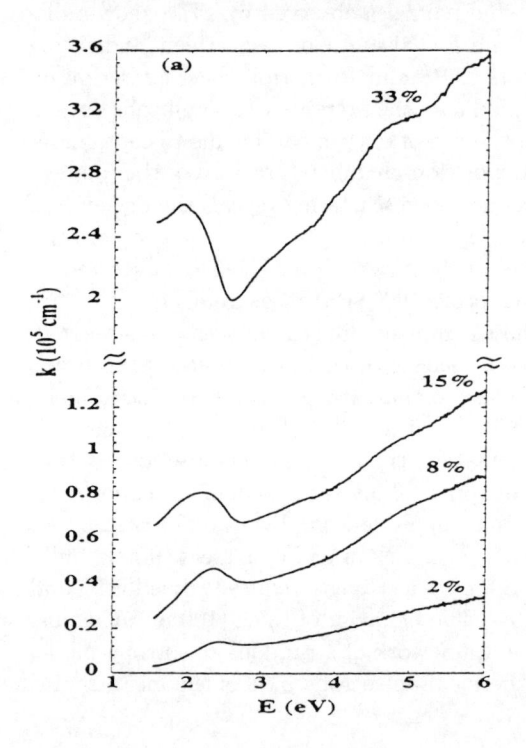

Fig. 5.23. Optical absorption spectra (absorption coefficient K versus photon energy E) obtained in 50 nm thick films of gold clusters (Au_{250}) in a LiF matrix. (From [5.20])

nature of an antiferromagnetic coupling of the clusters favored by the onset of a superexchange mechanism.

Deviations from a bulk-like behavior is shown by Fe clusters highly diluted in an Ag matrix produced by sputter-gas aggregation [5.96], or codeposition of cluster beams from a plasma aggregation source [5.97, 5.98]. Clusters with a size distribution centered around 1.1 nm show a mutual interaction and a superparamagnetic behavior above a temperature of 100 K. A significant dependence of the magnetic moment per atom is also observed in agreement to the reports on free iron clusters [5.99].

The control of size distribution and volumic fraction achievable with cluster beam co-deposition is of interest for the synthesis of optical nanostructures. Au and In clusters co-deposited with various matrix materials (SiO_x, LiF, MgF_2) maintain a narrow size distribution at concentrations lower than 10%, whereas at larger concentrations cluster coalescence is observed [5.20]. Optical absorption spectra show a dependence on the volumic fraction of the embedded particles (Fig. 5.23).

5.4 High-Energy Deposition

5.4.1 Implantation, Sputtering, Etching

As already suggested by simulations, the impact processes of energetic clusters are substantially different from those of monomer ions. Experiments conducted by Beuheler and co-workers [5.100, 5.101] have shown that kinetic energy is deposited at rates higher than predicted from classical two-body interaction stopping power theory. The sputtering yield is much higher compared to that observed with monomer ions; this is explained as being due to lateral sputtering effects. In Fig. 5.24 it is shown the angular dependence of

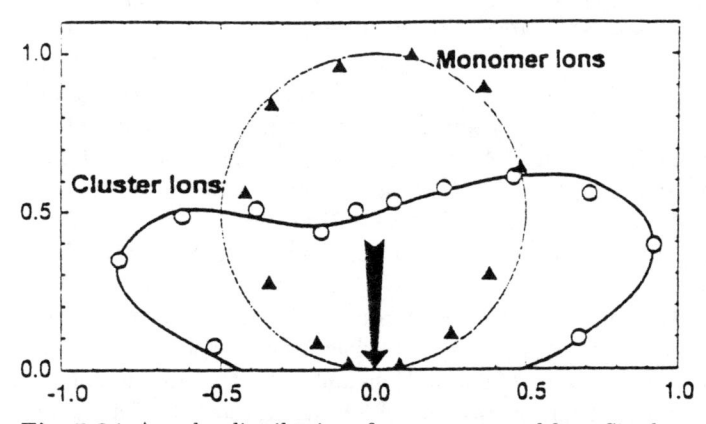

Fig. 5.24. Angular distribution of atoms sputtered from Si substrates due to normal incidence bombardment by 20 keV Ar_{3000} cluster ions and monomer Ar beam. (From [5.102])

sputtered atoms due to incident cluster ions showing different characteristics compared to the case of monomer ion bombardment. The mechanism causing this angular distribution may be the interaction of semi-spherical shock waves generated by cluster impact. Molecular dynamics simulations support this view [5.102].

Dissociation of cluster molecules due to the high energy impact, can also promote chemical reactions between the molecules and the target material, leading to the formation of volatile species and assisting the physical erosion of the target (Fig. 5.25) [5.103].

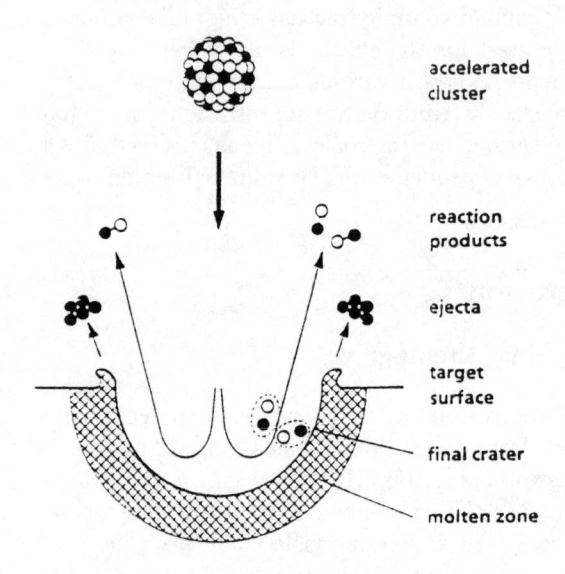

Fig. 5.25. Schematic view of crater formation by reactive sputtering induced by high-energy cluster impact. (From [5.103])

Yamada and co-workers [5.2, 5.104–5.107] have investigated the damage induced by gas cluster ion bombardment with particular attention to sputtering and implantation. Si (100) samples have been irradiated with 30 kV Ar clusters (mean size 150 atoms) ion beam of 335 pA corresponding to an estimated fluence of $0.7 \times 10^{15} \rightarrow 1.5 \times 10^{15}$ clusters cm^{-2}. Rutherford backscattering spectra show the formation of an amorphous Si layer of ∼5 nm thickness. A small portion of Ar atoms arriving on the surface (∼0.1%) is implanted corresponding to 1×10^{14} Ar atoms cm^{-2}. Bombardment with 30 keV, 1.3×10^{14} Ar monomers resulted in an implantation of 1.8×10^{16}. Strong smoothing effects induced by cluster beams on a Au film surface have also been observed.

When compared to monomer ion bombardment at normal incidence, cluster ion bombardment produces smoother surfaces: Cu surfaces bombarded with 20 keV Ar cluster to a dose of 8×10^{15} ions cm^{-2} show a reduction of surface roughness to 1.3 nm from the original of 5.8 nm. In the case of monomer Ar ion bombardment (1.2×10^{16} ions cm^{-2}), the obtained final roughness is 4.9 nm (Fig. 5.26).

5.4 High-Energy Deposition   157

Original surface (Ra=5.8nm)

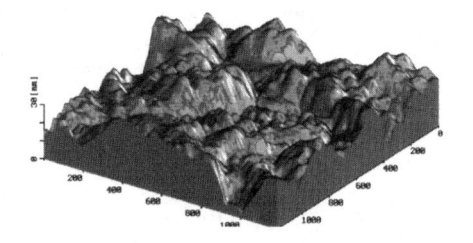

Ar monomer ion bombardment (Ra=4.9nm)
Dose=1.2×10^{16} ions/cm^2 at 20 keV

Ar cluster ion bombardment (Ra=1.3nm)
Dose of 8×10^{15} ions/cm^2 at 20 keV

Fig. 5.26. AFM images of Cu deposited Si substrate before and after sputtering. (From [5.102])

The smoothing effect is explained in terms of lateral sputtering, as suggested by numerical simulations [5.108], however, other mechanisms such as redeposition of ejected particles cannot be completely ruled out [5.109].

The dependence of sputtering yields upon the atomic number of the target material is reported in Fig. 5.27.

The etching of Si and SiO$_2$ by SF$_6$ cluster ions shows the ability of energetic clusters to perform anisotropic etching. Very high sputtering yields can be achieved with cluster beams that allow the realization of patterns which are difficult to prepare with conventional plasma etching techniques based on fluorine plasma (Fig. 5.28) [5.3].

Henkes and Krevet [5.110] have characterized the sputtering yield dependence on energy and current density for copper with a cluster beam of 1000 CO$_2$ of mean size. Increasing the cluster energy from 155 keV to 250 keV caused the copper sputter rate to increase three-fold, while the secondary electron/ion coefficient increases from 23 to 35. With the parameters used, the sputtering yield was not sensitive to current density. The dependence on target material for an ion current of 20 nA is reported in Table 5.1 [5.110].

The experimental erosion data do not show any correlation with parameters such as hardness or melting point of the material as would be the case if the sputtering by clusters were to work via cratering or evaporation respectively.

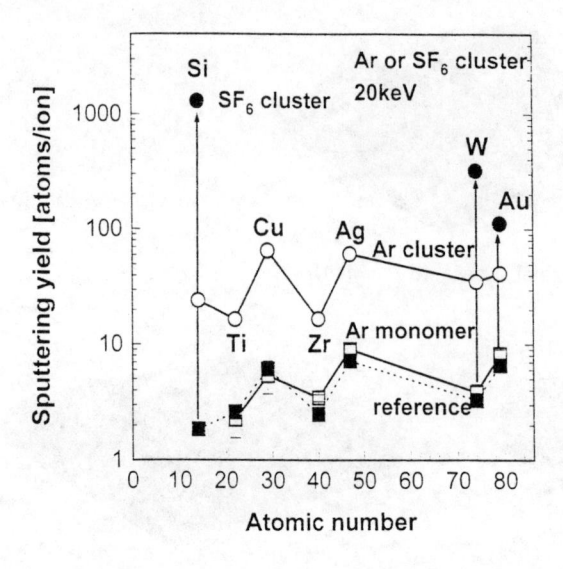

158   5. Thin Film Deposition and Surface Modification by Cluster Beams

Fig. 5.27. Dependence upon atomic number of sputtering yield due to 20 keV Ar cluster ions and monomer ions. Reactive sputtering yield of Si, W and Cu due to SF_6 cluster ion beam bombardment is also shown. (From [5.102])

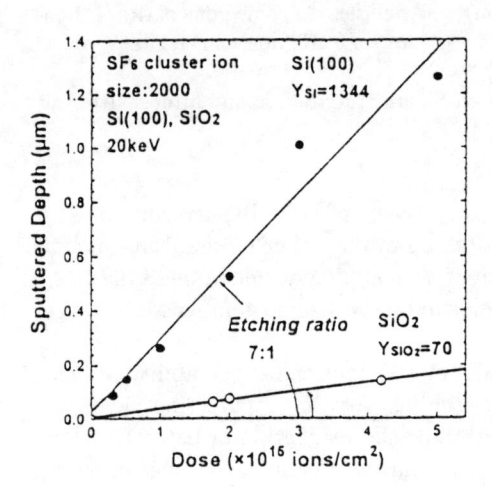

Fig. 5.28. Dose dependence of sputtered depth in Si and SiO_2 bombarded with SF_6 clusters. (From [5.102])

Table 5.1. Erosion rate of different materials using cluster ions. (From [5.110])

Materials	Erosion rate ($\mu m\,h^{-1}$)	Materials	Erosion rate ($\mu m\,h^{-1}$)
Tungsten	0.97	Copper	7.73
Aluminum	3.86	Magnesium	8.59
Nickel	4.42	Silver	11.6
Tanatalum	4.80	Silicon	19.3
Sapphire	4.83	PMMA	100.0
Diamond	6.65		

Damage, and in particular crater formation, by energetic ion cluster bombardment is reported for different surfaces (sapphire, silicon, graphite) [5.106, 5.107, 5.111]. For example, carbon clusters show an increased defect production for cluster sizes larger than six [5.112], fullerenes produce a higher defect density on silicon compared to carbon monomers [5.111].

In cluster ion beam processing the particles are used as an efficient means to transfer a high dose of energy onto a restricted surface region, this can be performed by using beams of clusters formed by gaseous species condensed by a supersonic expansion, the beam is then ionized and accelerated towards the target [5.113, 5.114]. In Fig. 5.29 a scheme of a typical apparatus used for surface modification is shown [5.114].

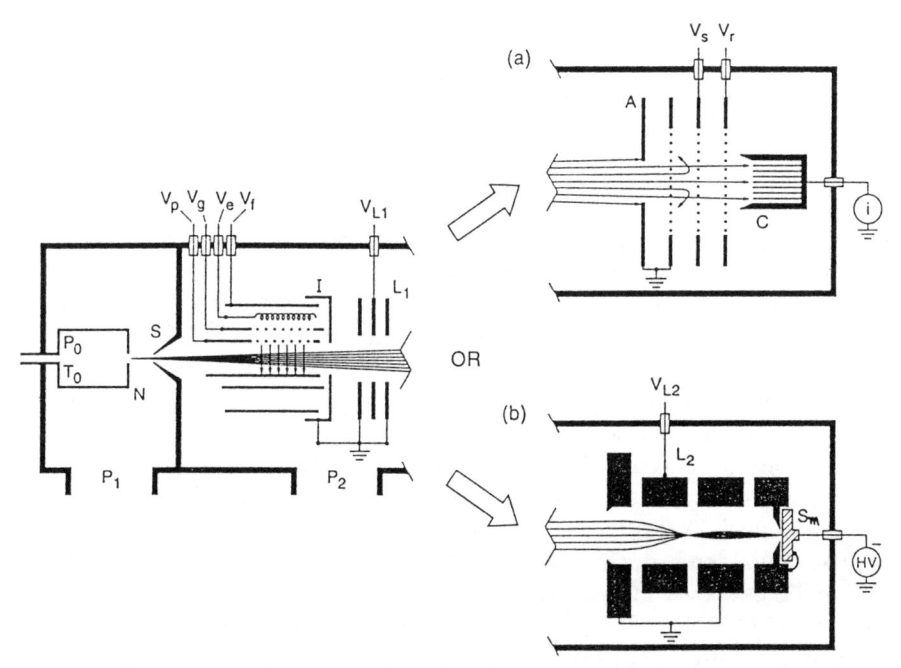

Fig. 5.29. Apparatus for surface modification with ion cluster beams, showing interchangeable electrode configurations: (**a**) analyzer with grids for beam characterization; (**b**) high voltage accelerating lens for sample exposure. (From [5.113])

Gas at the stagnation pressure P_0 and temperature T_0 expands through a nozzle N into a region maintained at a pressure $P \leq 10^{-3}$ Torr by a 25 cm diffusion pump. The central part of the beam is selected by a conical nozzle and enters a high vacuum region maintained at $P \leq 10^{-6}$ Torr. A cylindrical ionizer I, located at 7 cm from the skimmer generates an electron beam intersecting the neutral beam and ionizes the gas atoms and clusters. By adjusting the potential of the ionizer, only cluster ions are extracted and accelerated by a lens toward the substrate [5.113, 5.114].

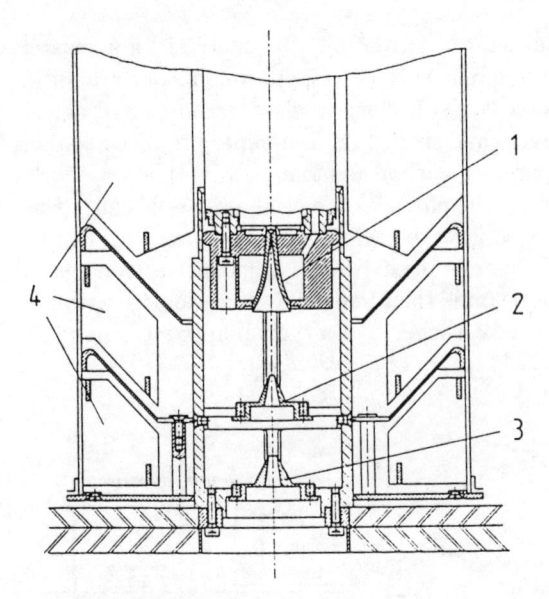

160   5. Thin Film Deposition and Surface Modification by Cluster Beams

Fig. 5.30. Cross-section of cluster beam source: 1–nozzle, 2–skimmer, 3–collimator, 4–cryopump. (From [5.116])

A high intensity cluster ion source used for micromachining [5.110, 5.115, 5.116] is shown in Fig. 5.30: the gas expands out of a trumpet-shaped nozzle kept at room temperature. A cluster beam is formed due to supersaturation caused by the expansion and it passes through two skimmers.

The pressure stages formed by the skimmers are cryopumped by copper fins cooled by liquid nitrogen. The clusters are ionized by electron impact in an ionizer integrated with an electrostatic extraction-focusing system. Since clusters are produced in a molecular beam, they all possess almost the same velocity. This implies a large kinetic energy spread, due the differences of masses, which makes focusing a major problem which is not encountered in conventional ion sources. A tungsten diaphragm (0.1 mm aperture) is placed behind the extraction electrode. The extraction voltage is adjusted so that the most probable size of the distribution is focused on this diaphragm. Smaller and larger clusters are out of focus and hit the diaphragm walls. Thus the beam can be partially monochromatized [5.116].

Beams of ionized CO_2 and SF_6 clusters, produced with this source, can be used for cluster impact lithography and polishing of surfaces. Diamond and cobalt-samarium ceramics can be structured using masks produced by the LIGA process [5.110], diamond, silicon and Pyrex glass have been microstructured with beams accelerated at 100 keV, showing very smooth eroded surfaces [5.103, 5.117].

5.4.2 Thin Film Formation

In 1972, Takagi and co-workers [5.6, 5.118, 5.119] proposed cluster beams for the growth of thin films. Metal clusters, produced by the expansion from a

5.4 High-Energy Deposition

crucible of a pure vapor, are ionized by electron impact and then accelerated at energies of several keV toward a substrate. A schematic diagram of the ionized cluster beam deposition (ICBD) apparatus is shown in Fig. 5.31. The material is heated in order to produce a vapor pressure inside the crucible $P_0 = 10^{-2}$ Torr up to several Torr [5.119]. The vapor thus formed is expanded through a nozzle with typical diameter $D = 0.5$–2.0 mm, the vacuum in the crucible chamber being in the range 10^{-7}–10^{-4} Torr. After ionization, clusters can be accelerated towards the substrate where their kinetic energy is converted into thermal, sputtering and implantation energies.

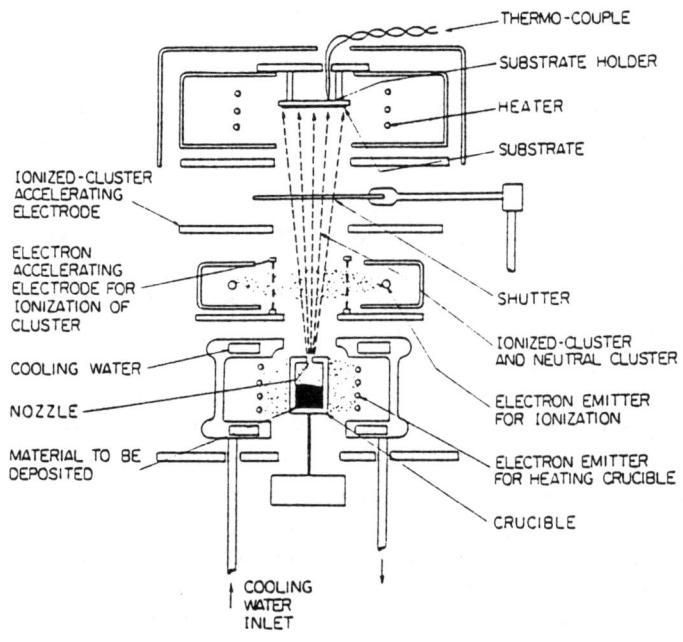

Fig. 5.31. Schematic view of a typical ICBD apparatus. (From [5.6])

One of the attractive features of using clusters instead of atoms is the way of depositing energy on the surface which should allow a substantial improvement in film adhesion and morphology while keeping the substrate temperature relatively low. The main parameters influencing the film growth are sticking coefficient, nucleation density and adatom mobilities. These parameters depend in turn on several factors such as deposition rate, substrate temperature, structural defects on the surface, contamination of the surface. The kinetic energy of all, or a fraction, of the impinging particles can strongly modify the above factors, and hence the growth parameters [5.120].

The distinctive features of ICBD as proposed by Takagi and Yamada are: improvement of adhesion by sputtering, etching and cleaning of the surface, localized deposition of energy enhancing diffusion and atom mobility, shallow ion implantation, control of nucleation and growth by deposition and migration of small aggregates [5.118]. A great improvement of the crystallinity of the film will result, in particular, from the enhanced mobility of the adatoms coming from momentum transfer and by thermal effects due to heating induced by the impact with the surface [5.121]. The kinetic energy of the impinging particles also affects the early stage of deposition and, in particular, the size distribution of the nucleation islands. Epitaxial layers, high quality thin films and nanostructures of several materials have been grown with ICBD [5.119, 5.122, 5.123].

Silicon epitaxial thin films have been deposited at a substrate temperature of 900 K at a reasonably high deposition rate using ICBD [5.119]. Smooth epitaxial aluminum films forming a very sharp interface, are also deposited on room temperature Si substrates despite the lattice mismatch of 25% [5.124]. A characterization of the first stages of the growth shows that at a temperature of 50 C diffusion distances from 13 µm to 17 µm are covered, the diffusion distance increases with the acceleration voltage. At high acceleration voltages, Auger spectroscopy and SEM shows that a transition from three-dimensional growth to a two-dimensional growth take place. The observed sharp Al/Si interface, with no interdiffusion, is attributed to the very small number of defects at the interface and in general to the lack of damage induced by ICB bombardment [5.124]. Using ICBD in a reactive atmosphere, thin films of oxides are also deposited [5.119].

ICBD processes have been optimized to obtain high fluxes and ionization efficiencies and applied to industrial processes for the fabrication of electronic devices and high quality mirror coatings [5.124, 5.125]. In particular films have been deposited in contact holes in DRAMs [5.126].

ICBD has achieved a considerable popularity and UHV-compatible cluster beam deposition systems are commercially available [5.127]. Although many groups have adopted this technique for high quality thin film deposition, quite surprisingly very few studies are devoted to the characterization of the cluster beams themselves. In particular, the cluster mass distribution before and after the electron impact ionization is inferred from source parameters and/or from indirect evidences (for example, characterization of the charged particle flux by retarding potential methods), instead of being directly measured by mass spectrometric techniques. This situation leads to misunderstandings and even to erroneous interpretation of the experimental results.

The derivation of the cluster mass distribution from velocity measurements should be considered with great care, since the observation of specific velocity regimes does not mean, ipso facto, that condensation has taken place during the expansion. Even the onset of a true supersonic regime should be checked carefully by measuring the beam velocities distribution.

According to the theoretical approach proposed by Hagena [5.128], as discussed in Chap. 2, the reduced condensation scaling parameter Γ^* can be used as a criterion for the onset of cluster formation. Standard ICBD conditions are characterized by $\Gamma^* \ll 200$ where clusters are not produced.

A time-of-flight mass spectroscopic analysis of a beam of germanium, produced by a typical ICBD source, shows the presence of clusters up to the tetramer with no evidence for larger clusters [5.129]. Analogous studies conducted on gallium and silver beams by retarding potential mass spectrometry and TOF, show that only a very small fraction of the beams is made of large clusters [5.130]. Experimental and computational analysis of the performance of an ionized cluster beam source also show that space charge effects in the ionization region strongly modify the energy distribution of the particles in the beam, thus invalidating the determination of mass distribution taken with a retarding potential spectrometer [5.131].

These results suggest that the ICBD technique is able to produce high quality thin films regardless of the presence of clusters, with a mechanism similar to low energy ion bombardment during vapor-phase deposition. Ion irradiation (20–100 eV) during thin film growth is shown to control the growth kinetics producing the same effects as those described for ICBD [5.120].

Hagena and co-workers [5.128, 5.132–5.135] describe systematically the conditions for metal cluster nucleation during supersonic expansion and perform thin film depositions with well characterized cluster beams. The apparatus used for cluster beam deposition is shown in Fig. 5.32. The cluster mass distributions are characterized by a retarding field detector [5.132].

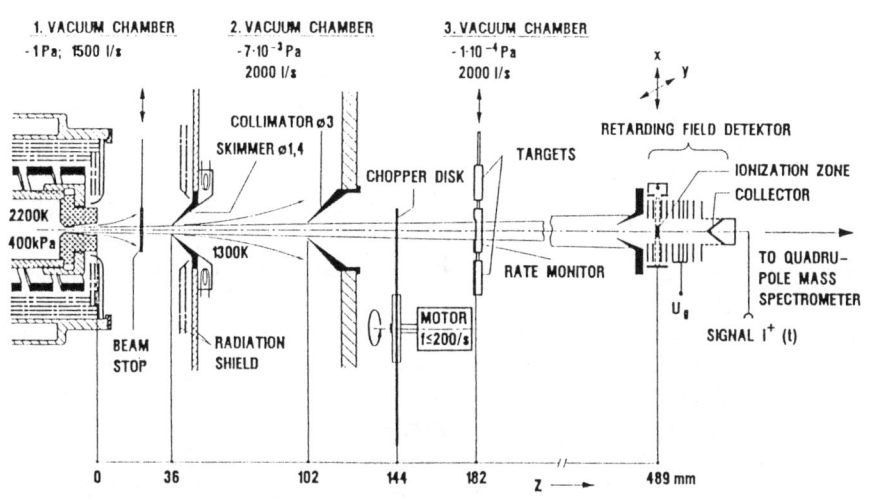

Fig. 5.32. Apparatus to produce and analyze silver cluster beams obtained from expanding silver vapor with argon as carrier gas. (From [5.128])

Silver cluster deposition rates up to $14\,\mathrm{nm\,s^{-1}}$ are measured at 0.78 m downstream the nozzle with a beam diameter of 94 mm. For typical expansion conditions requiring pumping speeds as low as $0.1\,\mathrm{m^3\,s^{-1}}$ at 25.5 Pa, silver neutral clusters can achieve kinetic energies of several hundred eV. Good quality thin films are obtained by using neutral cluster beams [5.128, 5.133].

Mg/Ar seeded beams produce clusters with a size distribution centered around 1700 monomers per aggregate whereas pure Mg expansions give 500 monomers per aggregate. The magnesium deposition rates for different types of beams are reported in Fig. 5.33 showing that a substantial improvement can be achieved compared to effusive flow.

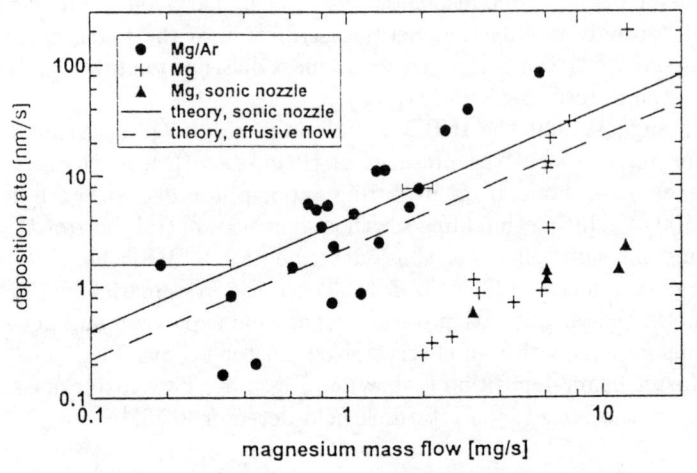

Fig. 5.33. Mg cluster beam deposition rate. (From [5.128])

Thick films (1000 nm) are deposited on room temperature glass and Si(111) substrates using Ar seeded and pure Mg beams. The films show a preferential orientation of (002) planes parallel to the substrate. The beam parameters affecting the final morphology of the films seem to be the deposition rate and the cluster mass distribution. The same film morphology is obtained by using seeded and pure beams with the other parameters constant [5.132].

Cluster beams from high temperature nozzle sources can be successfully used for metallization of microstructured surfaces. Metallic films with good crystalline properties can duplicate the microstructure of the surfaces up to an incidence angle of 80° [5.133, 5.134].

Metallization and hard coating deposition can also be obtained by using a magnetron plasma cluster source as described in Chap. 3 [5.136–5.138]

Compared to conventional gas-aggregation sources this has the advantage of being operated with refractory materials. It can also work in the reactive sputtering mode to produce, for example, nitride clusters. Varying the particle residence times in the aggregation chamber by regulating the quenching gas flow rate and/or the sputtering unit–nozzle distance, the cluster size distribution can be controlled. Ionized particles are efficiently produced without any post-expansion electron impact ionization. TOF analysis of the beam shows the production of ion clusters of several hundreds of monomer. By accelerating the clusters up to several keV towards room temperature substrates, strongly adhering and highly reflecting Mo and Cu films have been produced [5.136, 5.137]. TiN and $Ti_xAl_{1-x}N$ hard coatings have been also deposited at room temperature showing interesting structural and mechanical properties [5.138].

6. Outlook and Perspectives

In this final chapter we further discuss issues concerning cluster beam deposition in view of applications to the synthesis and processing of nanostructured materials.

A critical point resides in those aspects of cluster beam techniques (CBT) that make their transfer from academic laboratories to industry and production still challenging. The ultimate goal of cluster-based technology is the efficient production of monodisperse particles with controllable size and a precise spatial arrangement. Methods based on chemical routes allow the efficient production of systems formed by ordered arrays of monodisperse clusters, especially for II-VI semiconductors [6.1]. Cluster beams simply cannot compete on these grounds: deposition of size and energy selected clusters can be realized only for very small systems at very low deposition rates and the control over spatial arrangement of the particles is poor. Despite the fundamental interest in pursuing size-selected cluster beam deposition, applications of this technique are hampered by the extreme complexity of the experimental apparatus.

The synthesis and processing of nanostructured materials where only the mean cluster dimensions influence the functional and structural properties of the system can benefit from the use of cluster beams. Of course this technique is not convenient for the production of bulk materials, however nanostructured thin films and composite materials consisting of clusters embedded in transparent and polymeric matrices can be produced efficiently. These kind of materials are needed for optoelectronics, sensors, magnetic recording media, as well as for functional gradient coatings. The properties of these systems can be varied by controlling the mass and energy distribution of the cluster with a precision that does not require the use of beams monochromatic in mass and energy.

We believe that the major criteria that should guide a realistic evaluation on the applications of CBT are the following:

(i) the identification of a technological niche area where CBT offers solutions and performances unattainable with other techniques;

(ii) the possible CBT developments in view of scaleability and integration with different techniques of synthesis, analysis and processing.

6.1 Cluster Beam Processing of Surfaces

The exploration of cluster beams as a tool for material synthesis, processing and surface modifications is motivated by recent trends that are pushing the requirements in terms of dimensions, energy deposited, and processing temperatures towards the limits attainable with conventional ion implantation techniques [6.2].

We had several occasions, in the previous chapters, to discuss the advantages of cluster beams that we can summarize as follows: control of the kinetic energy of the aggregates and of the single components of a cluster in a range which is critical in the thin film growth process and for the activation of surface reactions; high directionality and high particle fluxes when supersonic beams are used (Fig. 6.1).

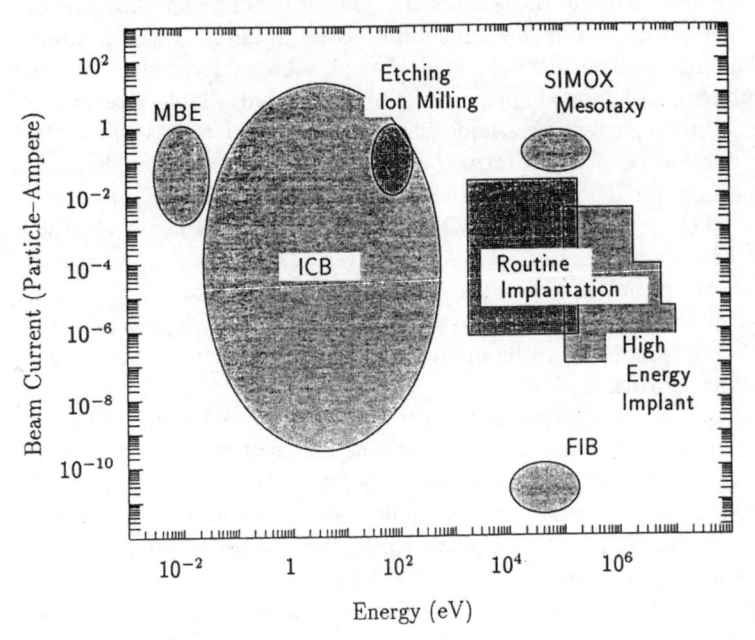

Fig. 6.1. Place of CBT among other ion beam techniques with respect to ion energy and beam current. (From [6.3])

As far as surface processing and implantation are concerned, cluster energy and flux are the main parameters that should be controlled since the mass distribution can be only roughly determined. A very promising direction is the etching and micro machining of surfaces carried out by reactive cluster beams. The sputtering and etching rates induced by cluster beams are competitive with more conventional methods such as atomic beam milling. The important advantage being that, coupled to the use of masks, CBT allows the microstructuring of refractory materials without any undesired implantation

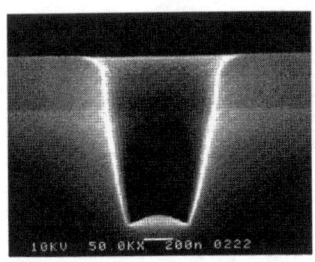

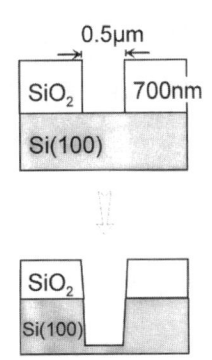

Fig. 6.2. SEM micrograph showing a cross-sectional view of Si substrate with a SiO$_2$ mask irradiated with SF$_6$ cluster ions. (From [6.2])

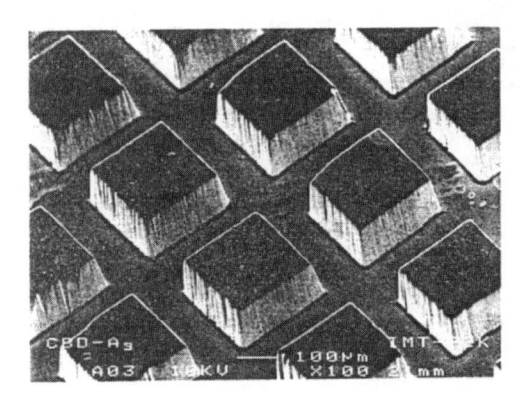

Fig. 6.3. Pyrex glass microstructured by CO$_2$ cluster impact. (From [6.5])

effects [6.4] while charge-up damage of the substrate is minimized due to the favorable mass to charge ratio of the clusters (Figs. 6.2 and 6.3).

The potential of CBT for the smoothing of metals, Si, CVD diamond and SiC has been demonstrated in the framework of large scale industrial programs [6.6]. Of particular relevance is the smoothing of high temperature superconductor (HTS) surfaces [6.7]. This method can be integrated into HTS circuit fabrication as a key step for planarization.

Another application of CBT of particular interest is the creation of shallow junctions by implantation of clusters and polyatomic molecules in semiconductors [6.8]. Very shallow p$^+$ junctions (less than 0.1 µm depth) can be formed by implanting decaborane (B$_{10}$H$_{14}$) into Si: a PMOSFET with 40 nm gate has been realized. The implantation of larger clusters should result in even shallower junctions [6.9].

Metallization and deposition of thin films and coatings, can also benefit from cluster beam deposition as already discussed in Chap. 5. It is worth remembering here that the high directionality and high intensity typical of continuous supersonic sources, make them particularly valuable for applications such as: metallization of isolating surfaces, filling the contact holes in

170   6. Outlook and Perspectives

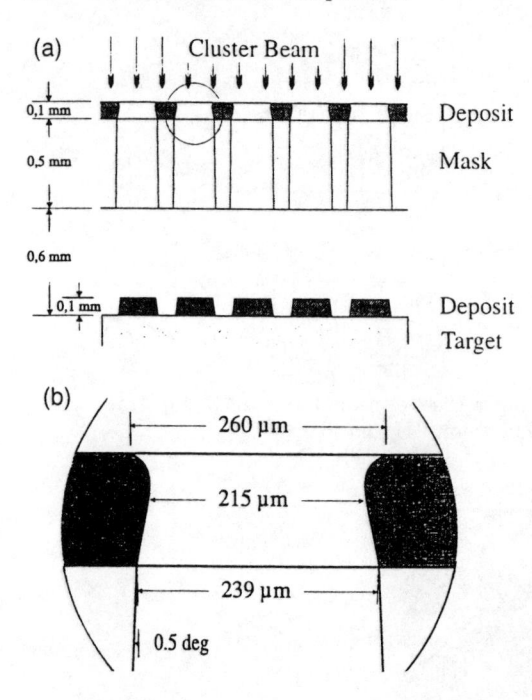

Fig. 6.4. Formation of microstructures using a grid mask. (a) Experimental arrangement of cluster beam, mask and substrate. (b) Detail of the silver deposit on the mask surface. (From [6.10])

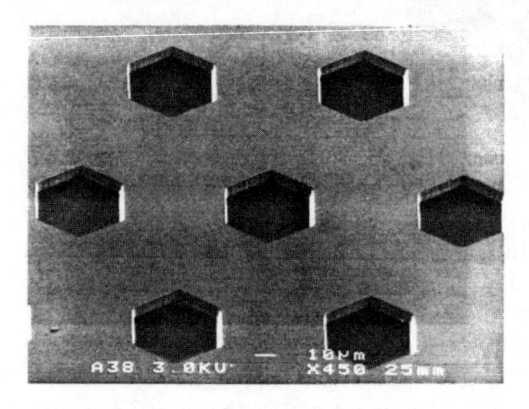

Fig. 6.5. SEM picture of silver bars 100 µm high produced with the mask shown in Fig. 6.4. (From [6.10])

combination with lift-off techniques and formation of thin foils of membranes with 3D microstructure [6.10]. Examples of these kinds of application are shown in Figs. 6.4 and 6.5.

These applications require a source of high brightness and stability. Unfortunately, supersonic sources that already meet these requirements can hardly be used with refractory materials. On the other hand, gas-aggregation sources, based on plasma discharges, give rise to encouraging performances and can be scaled to obtain very high intensities on large areas.

6.2 Nanostructured Materials Synthesis

Micromachining and surface processing in general require the use of clusters condensed from gaseous precursors by continuous supersonic expansions. This produces some experimental difficulties related to the handling of huge gas loads requiring efficient differential pumping systems and vacuum chambers. Moreover high efficiency ionization systems are necessary. On the other hand the process of condensation is controllable to a high degree of precision, supersonic sources of gas clusters are very stable and their scaleability depends only on the adequacy of the pumping speeds. The state of the art is that equipments based on gas cluster beams are nowadays routinely produced and used for industrial applications [6.11].

The synthesis of nanostructured materials is more demanding in terms of control of the cluster mass distribution, although it does not require cluster beams which are monochromatic in mass. The ideal cluster source for this application should be intense, stable and easy to operate. Moreover the ideal source should be compatible with UHV standards in order to deposit clusters in controlled conditions and to avoid contamination.

The choice between continuous and pulsed sources should take into account that the former can provide high duty cycles at the expense of the dimensions of the pumping apparatus. Pulsed sources seem to be a better candidate for the production of refractory cluster beams, moreover, they can be better coupled with UHV systems when fast growth rates are not critical.

Easy construction and operation are characteristic of pulsed sources that unfortunately do not correspond to a straightforward control on cluster beam parameters. The processes underlying cluster formation in pulsed plasma sources are poorly known on a microscopic level so that the control on the beam parameters is achieved on a phenomenological basis. The tuning of the cluster mass distribution by changing the source parameters is quite difficult and one should have the pragmatic attitude of carefully characterize the beam without inferring its properties from the behavior of similar sources, or from the characterization of pulsed valves alone. Stability is also a major problem at present. It should be noted that the actual generation of pulsed valves does not guarantee stable operation for sufficiently long times (hours) because of the presence of drifts. This is quite a complication in realizing the ideally simple and reliable cluster source.

The majority of pulsed plasma sources include a thermalization chamber that considerably alter the shape and duration of the gas delivered by the pulsed valve. This has several consequences: clusters have a long (milliseconds) residence time in the source, the pressure in the source is a function of time so that the beam characteristics also change. Generally speaking, these sources deliver beams that should be classified, as 'modulated' instead of 'pulsed'. In this regime the production of clusters is very efficient and the duty cycle larger than for a pure pulsed source. This of course requires a more efficient pumping system. From this point of view, cluster plasma sources for beam deposition are very similar to gas aggregation sources with a low duty cycle. This aspect should always be remembered in order to choose the right characterization protocol for this class of sources.

6.3 Perspectives

The overall picture is of a fast evolution, so that very promising future applications can be envisaged looking to recent developments that take advantage of specific properties intrinsic to cluster beams.

In particular, the deposition of thin films from supersonic molecular beams is recognized as an interesting alternative to conventional effusive beams in terms of intensity and collimation even applied to film growth of conventional semiconductors [6.12]. Supersonic cluster beams are very attractive in terms of beam collimation, thickness control, deposition rates for the synthesis of nanostructured layers where the degree of crystallinity can be controlled without thermal annealing [6.13]. Moreover, the possibility of using supersonic beams coupled with the standard thin film processing methods, opens a new route for the realization of devices containing nanostructured materials as integrated parts [6.14].

The synthesis of thin films and devices with integrated parts from precursors with typical dimensions of a few nanometers allows the introduction of structural and functional properties related to nanometer-scale architectures. This approach looks particularly promising for the integration of systems where an interplay between electronic structure and local geometric structure is important, as in the case of field emission, magnetic recording, non linear optics, energy storage and where objects characterized by different scales of dimensions are hierarchically organized [6.15].

A new challenging direction is also the application of free jets to the deposition of organic aggregates (molecules and oligomers). Key features are the control of the kinetic energy and the high directionality that make supersonic beams ideally suited to the ultra clean environments of growth chambers with *in situ* characterization. Organic molecular beam epitaxy is already becoming an alternative to chemical deposition to improve quality of films via a better control of the growth process. Recent experiments with seeded supersonic beams show that is possible to control the alignments (steric effect) of elongated molecules (oligo-thyophenes in particular) along the direction of the beam [6.16]. Such a feature coupled to the "tuning" of kinetic energy in a range (from one to several tens of eV) where the threshold for the activation of surface chemical processes lay, can further improve the quality of the growth extending the addressable surfaces and structures.

This shows how interconnected are the fundamental research and the developments of new applications: the difficulties, still numerous, to produce an industrial impact based on cluster beam methods should not cause one to overlook the large amount of work of characterization that is strongly needed.

A microscopic understanding of the role of precursors, beam parameters, first stages of growth, cluster coalescence, degree of cluster coalescence and hierarchical organization on the final structure of the film require the combination of an accurate characterization of the beam and of surface science characterization techniques in controlled conditions. This, in turn, requires an experimental approach that couples expertise from solid state-physics and molecular and atomic physics (gas phase) and the development of dedicated experimental apparatus, as shown schematically in Fig. 6.6, based on this interdisciplinary approach [6.17].

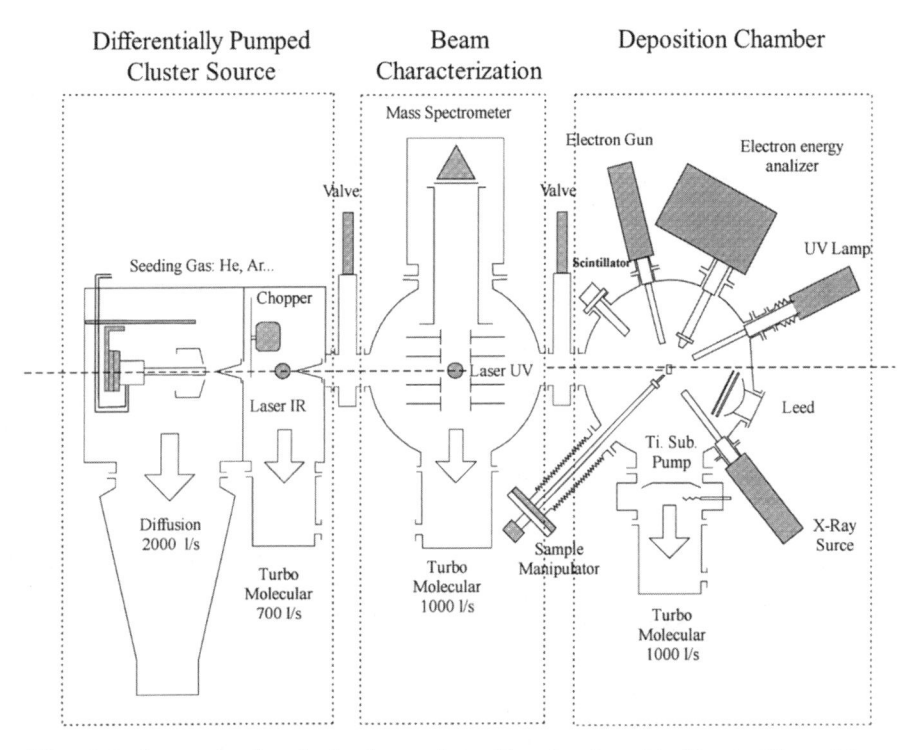

Fig. 6.6. Apparatus for cluster beam deposition developed in Trento. It combines characterization of the clusters in the beam to the study of the growth processes starting from the surface modifications and precursors. A cluster configuration of the apparatus and a fast entry lock allow the coupling with other sections, giving access to different kinds of analysis and processing

Appendix

Table A.1. List of some relevant properties of materials useful in the design of cluster sources and of interest as targets

Element	Melting point (K)	Boiling point (K)	Temperature for 1.5×10^{-2} mbar vapor (K)	Compatibility	Hazards
Aluminum (Al)	932	2740	1493	Reacts with ceramics, wets all materials, creeps, alloys with W	Flammable powder
Arsenic As_4 As_2	1093	883	583	Dimer vapor C pyrolized at 1473 K	Suspected carcinogen, highly toxic
Barium	983	2043	963	Reacts with ceramics, wets metals	Flammable solid
Beryllium	1556	2723	1473		Carcenogenic suspect highly toxic
Bismuth Bi_2 Bi	544	1803	948	Wets chromel Mo pyrolized at 1670 K	Flammable powder, harmful vapor
Cadmium	594	1040	533	Damages vacuum system, low sticking	Toxic, flammable powder, suspected carcinogen
Caesium	301	942	423	Explosive in air, low sticking, creeps damages detectors	
Calcium	1123	1713	873		Flammable powder
Chromium	2123	2873	1673		Toxic
Cobalt	1768	3173	1803	Alloys with refractory metals	Toxic flammable powder, suspected carcinogen

Element	Melting point (K)	Boiling point (K)	Temperature for 1.5×10^{-2} mbar vapor (K)	Compatibility	Hazards
Copper	1356	2853	1523		Flammable powder
Disprosium	1685	2573	1403		Flammable powder
Erbium	1803	2873	1523		Flammable powder
Europium	1099	1713	893		
Gadolinium	1585	2973	1873	alloys with Ta	
Gallium	303	2523	1323	Alloys with refractory metals	
Germanium	1223	3153	1673	Wets Ta and Mo	Flammable powder
Gold	1336	2933	1693	Wets Mo	
Holmium	1773	2573	1453		Flammable powder
Indium	429	2273	1213	Wets Cu, W	Toxic, flammable powder
Iron	1812	3173	1853	Alloys with all refractory metals	Flammable powder
Lead	600	2023	993		Toxic, flammable powder
Lithium	453	1603	808	Alloys with stainless steel	Flammable solid
Magnesium	923	1373	696		Flammable solid
Manganese	1523	2373	1233	Wets refractory metals	Flammable powder
Mercury	234.2	629.7	321		Highly toxic, harmful vapor
Molybdenum	2893	4873	2773	Sublimes	
Neodymium	1297	3443	1623		Flammable powder
Nickel	1726	3093	1773	Alloys with refractory metals	Suspected carcinogen
Palladium	1825	3413	1753		
Platinum	2045	4100	2373	Alloys with refractory metals	Flammable powder
Potassium	336.7	1047	487		Flammable solid
Praseodymium	1204	3785	1673		Flammable powder

Element	Melting point (K)	Boiling point (K)	Temperature for 1.5×10^{-2} mbar vapor (K)	Compatibility	Hazards
Rubidium	312	961	448	Explosive in air	Flammable solid
Samarium	1350	2064	1013	Sublimes	Flammable powder
Scandium	1814	3104	1653	Low alloying with W	
Selenium	490	958	563	Clusters easily, damages vacuum system, explosive in air	Toxic
Silicon	1683	2628	1893		Flammable powder
Silver	1235	2485	1303		Flammable powder
Sodium	371	1156	563		Flammable solid
Strontium	1042	1657	803	Wets refractory metals, no alloying	
Tantalum	3269	5700	3323		
Tellurium	722	1263	643	Clusters, wets metals, no alloying	Flammable solid, highly toxic
Terbium	1629	3396	1803		Flammable powder
Thallium	577	1733	903	Wets metals, no alloying	Toxic
Thorium	2023	5065	2673		
Thulium	1818	2220	1113	Sublimes	Flammable powder
Tin	505	2543	1523		Flammable powder
Titanium	1933	3560	2013	Reacts with refractory metals	
Tungsten	3683	5933	3523	Sublimes	Flammable powder
Vanadium	2163	3653	2123	Sublimes Wets Mo, no alloying	
Ytterbium	1092	1467	743	Sublimes	Flammable powder
Yttrium	1825	3611	1923		Flammable powder
Zinc	693	1180	613	Damage vacuum system, sublimes	Flammable powder
Zirconium	2125	4650	2723	Wets W with low alloying	

Table A.2. Some useful properties of ceramic materials of interest in source design

Material	dc resistance Ω cm		Coefficient of thermal expansion $\times 10^{-6}$		Melting Point (K)	Temperature for 1.5×10^{-5} mbar
	Temp. (K)	Value	Temp. (K)	Value		vapor (K)
Al_2O_3	1100	1×10^8	300–1100	8.5	2320	2040
	1370	1×10^6				
	2150	22				
BeO	900	3.5×10^8	300–1100	9.2	2790	2020
	1370	5.2×10^6				
ZrO_2	615	2.1×10^6	273–1673	5.0	2575	2570
	975	2.3×10^3				
	2270	0.59				
MgO	1075	3×10^8	273–1773	16	3075	1500
	1875	1×10^4				
ThO_2	825	2.6×10^7	300–1075	9.5	3300	2244
	1240	3.8×10^1				
BN		1×10^{15}			> 3270	3070
MACOR	300	$> 10^{16}$	300–1075	12.6		
SHAPAL-M	300	1×10^{12}	300–1075	5.1		

References

Chapter 1

1.1 W.A. de Heer: Rev. Mod. Phys. 65, 611 (1993)
1.2 H. Haberland (ed.): *Clusters of Atoms and Molecules*, Springer-Verlag, Berlin-Heidelberg, 1994
1.3 Proc. of the 8th International Symp. on Small Particles and Inorganic Clusters, ISSPIC 8, Z. Phys. D **40**, No 1–4, (1997)

Chapter 2

2.1 O.F. Hagena: in, *Molecular Beams and Low Density Gas Dynamics*, P.P. Wegener (ed.), (Dekker, New York, 1974)
2.2 H. Haberland (ed.): *Cluster of Atoms and Molecules*, (Springer-Verlag, Berlin-Heidelberg, 1994)
2.3 G. Scoles (ed.): *Atomic and Molecular Beam Methods*, (Oxford University Press, Oxford, 1988)
2.4 N. F. Ramsey: *Molecular Beams*, (Oxford University Press, Oxford, 1955)
2.5 P. Clausing: Ann. Phys. **12**, 961 (1932)
2.6 J.A. Giordamaine, T.C. Wang: J. Appl. Phys. **31**, 463 (1960)
2.7 D.M. Murphy: J. Vac. Sci. Technol. A**7**, 3075 (1989)
2.8 C.B. Lucas: Vacuum **23**, 395 (1972)
2.9 K.J. Ross: Rev. Sci. Instrum. **66**, 4409 (1995)
2.10 D.R. Miller: in ref [2.3], vol 1, p. 15
2.11 Beam Dynamics Inc, 708 East 56th Street - Minneapolis, Minnesota 55417 USA
2.12 M. Zucrow, J. Hoffman, *Gas Dynamics*, vols. I and II, Wiley, New York, 1976
2.13 H. Ashkenas, F.S. Sherman:*4-th Rarefied Gas Dynamics Symposium*, **2**, 784 (1966)
2.14 J.B. Anderson: AIAA J. **10**, 112 (1972)
2.15 J.P. Toennies, K. Winkelmann: J. Chem. Phys. **66**, 3965 (1977)
2.16 H. Beijering, N. Verster, Physica C **111**, 327 (1981)
2.17 P. Poulsen, D.R. Miller: *10-th Rarefied Gas Dynamics Symposium*, **2**, 899 (1977)
2.18 P.C. Waterman, S.A. Stern: J. Chem. Phys. **31**, 405 (1959)
2.19 E.W. Becker, K. Bier, W. Bier: Z. Naturforsch. A **16**, 12 (1961)
2.20 V.H. Reis, J.B. Fenn: J. Chem. Phys. **39**, 3240 (1963)
2.21 G.W. Israel, S.K. Friedlander: J. Colloid Interface Sci. **24**, 330 (1967)
2.22 J. Fernandez De La Mora, B.L. Halpern, J.A. Wilson: J. Fluid Mech. **149**, 217 (1984)

2.23 W.R. Gentry: in ref [2.3], vol 1, p 54
2.24 M. Volmer, A. Weber: Z. Phys. Chem. **119**, 277 (1925)
2.25 J.B. Zeldovich: Acta Physicochim., U.R.S.S. **18**, 1 (1943)
2.26 J. Frenkel, *Theory of Liquids,* (Oxford Univeristy Press, Oxford, 1946)
2.27 D. Turnbull: J. Chem. Phys. **18**, 198 (1950)
2.28 D. Turnbull: J. Chem. Phys. **20**, 411 (1952)
2.29 A. Laaksonen, V. Talanquer, D.W. Oxtoby: Annu. Rev. Phys. Chem. **46**, 489 (1995)
2.30 D.W. Oxtoby: J. Phys.: Condens. Matter **4**, 7627 (1992)
2.31 D.T. Wu: Solid State Phys. **50**, 37 (1997).
2.32 J.W. Christian: *The Theory of Transformation in Metals and Alloys,* 2nd ed., Pergamon Press, Oxford, 1975
2.33 R.A. Sigsbee: in *Nucleation,* A.C. Zettlemoyer (ed.), p.151, (Dekker, New York, 1969)
2.34 R. Rechsteiner, J.D. Ganière: Surf. Sci. **106**, 125 (1981)
2.35 O. Abraham, S.S. Kim, G.D. Stein: J. Chem. Phys. **75**, 402 (1981), and references therein.
2.36 G.D. Stein: Surf. Sci. **156**, 44 (1985)
2.37 B.K. Rao, B.M. Smirnov: Phys. Scr. **46,** 439 (1997)
2.38 R.E. Smalley: Laser Chem. **2**, 167 (1991)
2.39 J. Khatoun, M. Mostafavi, J. Ambland, J. Belloni: Z. Phys D **34**, 47 (1995)
2.40 O.F. Hagena: Surf. Sci. **106**, 101 (1981)
2.41 O.F. Hagena: Phys. Fluids **17**, 894 (1974)
2.42 O.F. Hagena, W. Obert: J. Chem. Phys. **56**, 1793 (1972)
2.43 H.P. Birkhofer, H. Haberland, M. Winterer, D.R. Worshop: Ber. Bunsenges. Phys. Chem. **88**, 207 (1984)
2.44 O.F. Hagena: Z. Phys. D**4**, 291 (1987)
2.45 O.F. Hagena: in *Rarefied Gas Dynamics,* H. Oguchi (ed.), (University of Tokyo Press, Tokyo, 1984)
2.46 R. Weiel: Z. Phys D**27**, 89 (1993)

Chapter 3

3.1 K.J. Ross, B. Sonntag: Rev. Sci. Instrum. **66**, 4409 (1995)
3.2 C.J. Smithells: *Metals Reference Book,* (Butterworth, London, 1962)
3.3 *Handbook of Chemistry and Physics, 57th ed,* (Chemical Rubber, Boca Raton, 1983)
3.4 R.J. Lewis (ed.): *Registry of Toxic Effects of Chemical Substances,* (U.S. Department of Health, Education and Welfare, Washington D.C., 1979)
3.5 W.H. Kohl: *Materials and Techniques for Electron Tubes,* (Chapman and Hall, London, 1960)
3.6 *Balzers Coatings Materials Sputtering Targets,* Evaporation Sources 1990/1992
3.7 A. Hausmann, B. Kammerling, H. Kossmann, V. Schmidt: Phys. Rev. Lett. **23**, 2669 (1988)
3.8 J.M. Dyke, A. Morris, J.C.H. Stevens: Chem. Phys. **102**, 29 (1986)
3.9 J.M. Dyke, S. Elbel, A. Morris, J.C.H. Stevens: J. Chem. Soc. Faraday Trans. **82**, 637 (1986)
3.10 M.O. Krause, A. Svensson, A. Fahlmann, T.A. Carlson, F. Cerrina: Z. Phys. D**2**, 327 (1986)
3.11 J.M. Dyke, N.K. Fayad, G.D. Josland, A. Morris: Chem. Phys. **67**, 245 (1982)
3.12 J.M. Dyke, N.K. Fayad, A. Morris, I.R. Trickle: J. Phys **B 12**, 2985 (1979)

3.13 V.E. Bondybey, G.P. Schwartz, J.E. Griffiths: J. Mol. Spectrosc. **89**, 328 (1981)
3.14 L.I. Maissel, R. Gland (eds.): *Handbook of Thin Films Technology*, (McGraw-Hill, New York, 1970)
3.15 P.D. Johnson: J. Am. Ceram. Soc. **33**, 168 (1950)
3.16 R. Kieffer, F. Benesovsky: Metalurgia **58**, 119 (1958)
3.17 W.H. Crumley, J.S. Hayden, J.L. Gole: J. Chem. Phys. **84**, 5250 (1986)
3.18 J.S. Hayden, R. Woodward, J.L. Gole: J. Phys. Chem. **90**, 1799 (1986)
3.19 D. Bulgin, J. Dyke, J. Goodfellow, N. Jonathan, E. Lee, A.J. Morris: Electron Spectrosc. **12**, 67 (1977).
3.20 D.B. Chrisey, G.K. Hubler (eds): *Pulsed Laser Deposition of Thin Films*, (Wiley, New York, 1994)
3.21 J.C. Miller (ed): *Laser Ablation*, Springer Ser. Mater. Sci., vol. 28, (Springer, Berlin, Heidelberg, 1994)
3.22 M. von Allmen (ed.): *Laser-Beam Interactions with Materials*, Springer Ser. Mater. Sci., vol. 2, (Springer, Berlin, Heidelberg, 1987)
3.23 R.E. Smalley: Laser Chem. **2**, 167 (1983)
3.24 P. Milani, W.A. deHeer: Rev. Sci. Instrum. **61**, 1835 (1990)
3.25 M. Samy El-Shall, W. Slack, W. Vann, D. Kane, D. Hanley: J. Phys. Chem. **98**, 3067 (1994)
3.26 J.M. Ballestreros, R. Serna, J. Solis, C.N. Afonso, A.K. Petford-Long, D.H. Osborne, R.F. Haglund, Jr.: Appl. Phys. Lett. **71**, 2445 (1997)
3.27 J. Cheung, J. Horwitz: MRS Bull. 30 (1992)
3.28 A. Peterlongo, A. Miotello, R. Kelly: Phys. Rev. E **50**, 4716 (1994)
3.29 W. Kautek, B. Roas, L. Schultz: Thin Solid Films **191**, 317 (1990)
3.30 J.T. Cheung, H. Sankur, Crit. Rev. Solid State Mater. Sci. **15**, 63 (1988)
3.31 C.L. Chan, J. Mazumder: J. Appl. Phys. **62**, 4579 (1987)
3.32 E. van de Riet, C.J.C.M. Nillesen, J. Dieleman: J. Appl. Phys. **74**, 2008 (1993)
3.33 J.F. Ready: *Industrial Applications of Lasers*, (Academic Press, New York, 1976)
3.34 H.J. Freund, S.H. Bauer: J. Phys. Chem. **81**, 994 (1977)
3.35 M.A. Lieberman, A.J. Lichtemberg: *Principles of Plasma Discharges and Materials Processing*, (Wiley, New York, 1994)
3.36 Y.P. Raizer: *Gas Discharge Physics*, (Springer, Berlin, Heidelberg, 1991)
3.37 G. Disatnik, R.L. Boxman, S. Goldsmith: IEEE Trans. Plasma Sci. **PS-15**, 520 (1987)
3.38 S. Anders, A. Anders: IEEE Trans. Plasma Sci. **19**, 20 (1991)
3.39 J. Vyskočil, J. Musil: Surf. Coat. Technol. **43/44**, 299 (1990)
3.40 H. Randhawa, P.C. Johnson: Surf. Coat. Technol. **31**, 303 (1987)
3.41 R.L. Boxman, S. Goldsmith: Surf. Coat. Technol. **33**, 153 (1987)
3.42 T. Utsumi: Appl. Phys. Lett. **18**, 218 (1971)
3.43 H.C. Miller: IEEE Trans. Plasma Sci. **PS-13**, 242 (1985)
3.44 S. Shalev, R.L. Boxman, S. Goldsmith: J. Appl. Phys. **58**, 2503 (1985)
3.45 B. Gellert, E. Schade, E. Dullni: IEEE Trans. Plasma Sci. **PS-15**, 545 (1987)
3.46 C.W. Kimblin: J. Appl. Phys. **44**, 3014 (1973)
3.47 C.W. Kimblin: J. Appl. Phys. **45**, 5235 (1974)
3.48 M.G. Drouet, J.-L. Meunier: IEEE Trans. Plasma Sci. **PS-13**, 285 (1985)
3.49 J.-L. Meunier, M.G. Drouet: IEEE Trans. Plasma Sci. **PS-15**, 515 (1987)
3.50 J.-L. Meunier: IEEE Trans. Plasma Sci. **18**, 904 (1990)
3.51 L.D. Landau, E.M. Lifshits: *Statistical Physics*, vol. 1, (Pergamon Press, Oxford, 1980)
3.52 O. Hagena: Surf. Sci. **106**, 101 (1981)
3.53 O.F. Hagena: Rev. Sci. Instrum. **63**, 2374 (1991)

3.54 K. Gingerich, I. Shim, S. Gupta, J. Kingcade, Jr: Surf. Sci. **156**, 495 (1985)
3.55 T. Scheuring, K. Weil: Surf. Sci. **156,** 457 (1985)
3.56 T.P. Martin: Angew. Chem. Int. Ed. Engl. **25**, 197 (1986)
3.57 K. Sattler, J. Mühlbach and E. Recknagel: Phys. Rev. Lett. **45**, 821 (1980).
3.58 G. Fuchs, P. Melinon, F. Santos Aires, M. Treilleux, B. Cabaud and A. Hoareau: Phys. Rev. B **44**, 3926 (1991).
3.59 G. Fuchs, P. Melinon, Y. Yan, B. Cabaud, A. Hoareau, M. Treilleux, V. Paillard: Suppl. to Z. Phys. D **26**, S 249 (1993).
3.60 U. Zimmerman, N. Malinowski, U. Näher, S. Frank, T.P. Martin: Z. Phys. D **31**, 85 (1994)
3.61 I.M. Goldby, B. von Issendorff, L. Kuipers, R.E. Palmer: Rev. Sci Instrum. **68**, 3327 (1997)
3.62 S.H. Baker, S.C. Thorton, A.M. Keen, T.I. Preston, C: Norris, K.W. Edmonds and C. Binns: Rev. Sci. Instrum. **68**, 1853 (1997)
3.63 H. Hahn, R.S. Averback: J. Appl. Phys. **67**, 1113 (1990)
3.64 H. Haberland, M. Karrais, M. Mall, Y. Thurner: J. Vac. Sci. Technol. A **10**, 3266 (1992)
3.65 H. Haberland, M. Mall, M. Moseler, Y. Qiang, T. Reiners, Y. Thurner: J. Vac. Sci. Technol. A **12**, 2925 (1994)
3.66 S. Yamamuro, M. Sakurai, T.J. Konno, K. Sumiyama, K. Suzuki: in *Similarities and Differences between Atomic Nuclei and Clusters*, Y. Abe, I. Arai, S.M. Lee, K. Yabana (eds.), (AIP Conf. Proc., vol 416, Woodbury N.Y., 1998)
3.67 S. Yamamuro, K. Sumiyama, M. Sakurai, K. Suzuki: Supramolec. Sci. **5**, 239 (1998)
3.68 G.D. Stein: Surf. Sci. **156**, 44 (1985)
3.69 E.W. Becker, K. Bier, W. Henkes: Z. Phys. **146**, 333 (1956)
3.70 O.F. Hagena, W. Obert: J. Chem. Phys., **56**, 1793 (1972)
3.71 G.D. Stein, J.A. Armstrong: J. Chem. Phys. **58**, 1999 (1973)
3.72 J. Gspann: Surf. Sci., **106**, 219 (1981)
3.73 P. Gatz and O.F. Hagena: J. Vac. Sci. Technol. A **13**, 2128 (1995)
3.74 S. Christ, P.M. Sherman, D.R. Glass: AIAA J. **4**, 68 (1966)
3.75 M. Kappes, R. Kunz, E. Schumacher: Chem. Phys. Lett. **91**, 413 (1982)
3.76 O. Abraham, S. Kim, G.D. Stein: J. Chem. Phys., **75**, 402 (1981)
3.77 E.W. Schlag, H.L. Selzle: J. Chem-Soc. Faraday Trans. **86**, 2511 (1990)
3.78 P. Gatz, O.F. Hagena: Appl. Surf. Sci., **91**, 169 (1995)
3.79 A. Amirav: Comments At. Mol. Phys. **24**, 187 (1990)
3.80 F. Biasioli, A. Boschetti, E. Barborini, P. Piseri, P. Milani, S. Iannotta: Chem. Phys. Lett., **270**, 115 (1997)
3.81 G. Ciullo, F. Biasioli, R. Dwivedi, M. van Opbergen, A. Podestà, A. Boschetti, S. Iannotta: Proc. Symposium on Atomic and Surface Physics (SASP 98), (Going in Tirol, Austria, 1998)
3.82 O.F. Hagena: Z. Phys. D **20**, 425 (1991)
3.83 U. Heiz, A. Vayloyan, E. Schumacher: Rev. Sci. Instrum.,**68**, 3718 (1997)
3.84 K. Nagaya, T. Hayakawa, M. Yao, H. Endo: J. of Non-Cryst. Solids, **205**, 807 (1996)
3.85 W.R. Gentry: in *Atomic and Molecular Beam Methods*, G. Scoles (ed.), (Oxford University Press, Oxford 1988)
3.86 C.E. Otis, P.M. Johnson: Rev. Sci. Instrum. **51**, 1128 (1980)
3.87 D. Bassi, S Iannotta, S. Niccolini: Rev. Sci. Instrum. **52**, 8 (1981)
3.88 T.E. Adams, B.H. Rockney, R.J.S. Morrison, E.R. Grant: Rev. Sci. Instrum. **52**, 1469 (1981)
3.89 General Valve Corporation, Series 9

3.90 L. Abad, D. Bermejo, V.J. Herrero, J. Santos, I. Tanarro: Rev. Sci. Instrum. **66**, 3826 (1995)
3.91 M.R. Adriaens, W. Allison, B. Feuerbacher: J. Phys. E: Sci. Instrum. **14**, 1375 (1981)
3.92 A. Auerbach, R. McDiarmid: Rev. Sci. Instrum. **51**, 1273 (1980)
3.93 J.B. Cross, J.J. Valentini: Rev. Sci. Instrum. **53**, 38 (1982)
3.94 P. Andresen, M. Faubel, D. Haeusler, G. Kraft, H.-W. Luelf, J.G. Skofronick: Rev. Sci. Instrum. **56**, 2038 (1985)
3.95 D. Proch, T. Trickl: Rev. Sci. Instrum. **60**, 713 (1989)
3.96 W.R. Gentry, C.F. Giese: Rev. Sci. Instrum. **49**, 595 (1978)
3.97 L. Li, D.M. Lubman: Rev. Sci. Instrum. **60**, 499 (1989)
3.98 S.M. Senken, S.C. Deskin: Rev. Sci. Instrum. **68**, 4286 (1997)
3.99 J.P. Bucher, D.C. Douglass, P. Xia, L.A. Bloomfield: Rev. Sci. Instrum. **61**, 2374 (1990)
3.100 O.F. Hagena: Rev. Sci. Instrum. **62**, 2038 (1991)
3.101 T.G. Dietz, M.A. Duncan, D.E. Powers, R.E. Smalley: J. Chem. Phys. **74**, 6511 (1981)
3.102 M.E. Geusic, M.D. Morse, S.C. O'Brien, R.E. Smalley: Rev. Sci. Instrum. **56**, 2123 (1985)
3.103 P.J. Brucat, L.S. Zeng, C.L. Pettiette, S. Yang, R.E. Smalley: J. Chem. Phys. **84**, 3078 (1986)
3.104 J. Woenckhaus, J.A. Becker: Rev. Sci. Instrum. **65**, 2019 (1994)
3.105 M. Pellarin et al.: Chem Phys. Lett. **224**, 338 (1994)
3.106 P. Gangopadhyay, J.M. Lisy: Rev. Sci. Instrum. **62**, 502 (1991)
3.107 U. Heiz, F. Vanolli, L. Trento, W.-D. Schneider: Rev. Sci. Instrum. **68**, 1986 (1997)
3.108 H.V. Kroto, J.R. Heath, S.C. O'Brien, R.F. Curl, R.E. Smalley: Nature **318**, 162 (1985)
3.109 S. Maruyama, L.R. Anderson, R.E. Smalley: Rev. Sci. Instrum. **61**, 3686 (1990)
3.110 S. Schwyn, E. Garwin, A. Schmidt-Ott: J. Aerosol Sci. **19**, 639 (1988)
3.111 W. Kraetschmer, L.D. Lamb, K. Fostiroupolos, D.R. Huffman: Nature **347**, 354 (1990)
3.112 M.H. Teng, J.J. Host, J.-H. Hwang, B.R. Elliott, J.R. Weertman, T.O. Mason, V.P. Dravid, D.L. Johnson: J. Mater. Res. **10**, 233 (1995)
3.113 S.L. Girshick, C.-P. Chiu: Plasma Chem. Plasma. Process. **9**, 355 (1989)
3.114 W.A. Saunders, P.C. Sercel, R.B. Lee, H.A. Atwater, K.J. Vahala, R.C. Flagan, E.J. Escorcia-Aparcio: Appl. Phys. Lett. **63**, 1549 (1993)
3.115 H.D. Li, H.B. Yang, S. Yu, G.T. Zou, Y.D. Li, S.Y. Liu, S.R. Yang: Appl. Phys. Lett. **69**, 1285 (1996)
3.116 G. Ganteför, H.R. Siekmann, H.O. Lutz, K.H. Meiwes-Broer: Chem. Phys. Lett. **165**, 293 (1990)
3.117 H.R. Siekmann, Ch. Lüder, J. Faehrmann, H.O. Lutz, K.H. Meiwes-Broer: Z. Phys. D **20**, 417 (1991)
3.118 H.R. Siekmann, E. Holub-Krappe, Bu. Wrenger, Ch. Pettenkofer, K.H. Meiwes-Broer: Z. Phys. B **90**, 201 (1993)
3.119 C.Y. Cha, G. Ganteför, W. Eberhardt: Rev. Sci. Instrum. **63**, 5661 (1992)
3.120 M. Gausa, R. Kaschner, G. Seifert, J.H. Faehrmann, H.O. Lutz, K.H. Meiwes-Broer: J. Chem. Phys. **104**, 9719 (1996)
3.121 P. Milani, M. Ferretti, P. Piseri, C.E. Bottani, A. Ferrari, A. Li Bassi, G. Guizzetti, M. Patrini: J. Appl. Phys. **82**, 5793 (1997)
3.122 P. Milani, P. Piseri, C.E. Bottani, A. Li Bassi: in *Similarities and Differences between Atomic Nuclei and Clusters*, Y. Abe, I. Arai, S.M. Lee, K. Yabana (eds.), (AIP Conf. Proc., vol 416, Woodbury N.Y., 1998)

Chapter 4

4.1 W. Paul, M. Raether: Z. Phys. **140**, 262 (1955)
4.2 W. Paul, H.P. Reinhard, U. von Zahn: Z. Phys. **152**, 143 (1958)
4.3 F. Von Busch, W. Paul: Z. Phys. **164**, 581 (1961)
4.4 P.H. Dawson, *Quadrupole Mass Spectrometry and its Applications,* (Elsevier, Amsterdam, 1976)
4.5 D. Bassi: in *Atomic and Molecular Beam Methods,* G. Scoles (ed.), Vol. I, Chap. 8, (Oxford University Press, Oxford, 1988)
4.6 J.J. Tunstall, A.C.C. Voo, S. Taylor: Rap. Comm. Mass Spectr. **11**, 184 (1997)
4.7 A.C.C. Voo, R. Ng, J.J. Tunstall, S. Tylor: J. Vac. Sci. Technol A **15**, 2227 (1997)
4.8 W.M. Brubaker: in *Advances in Mass Spectrometry,* A. Quoule (ed.), (Elsevier, Amsterdam, 1968)
4.9 B.A. Collings, D.J. Douglas: Int. J. Mass Spectrom. Ion Process. **162**, 121 (1997)
4.10 S. Vajda, S. Wolf, T. Leisner, U. Busolt, L.H. Woste, D.J. Wales: J. Chem. Phys. **107**, 3492 (1997)
4.11 R.L. Wagner, W.D. Vann, A.W. Castleman: Rev. Sci. Instrum. **68**, 3010 (1997)
4.12 A.N. Zavilopulo, A.L. Dolgin: Nucl. Instrum. Methods Phys. Res. B **126**, 305 (1997)
4.13 see, for example: Int. J. Mass Spectrom. Ion Process. **131** (1994)
4.14 W.C. Wiley, I.H. McLaren: Rev. Sci. Instrum. **26**, 1150 (1955)
4.15 T. Bergmann, T.P. Martin, H. Schaber: Rev. Sci. Instrum. **60**, 347 (1989); ibid. **61**, 2592 (1990); T. Bergmann, H. Goelich, T.P. Martin, H. Schaber, G. Malegiannakis: Rev. Sci. Instrum. **61**, 2585 (1990)
4.16 U. Boesl, R. Weinkauf, C. Weickhardt, E.W. Schlag: Int. J. Mass Spectrom. Ion. Process. **131**, 87 (1994)
4.17 W.A. deHeer, P. Milani: Rev. Sci. Instrum. **62**, 670 (1991)
4.18 G. Sanzone: Rev. Sci. Instrum. **41**, 741 (1970)
4.19 V.I. Karataev, B.A. Mamyrin, D.V. Shmikk: Sov. Phys. -Tech. Phys. **16**, 1177 (1972)
4.20 D.M. Lubman, R.M. Jordan: Rev. Sci. Instrum. **56**, 373 (1985)
4.21 R.B. Opsal, K.G. Owens, J.P. Reilly: Anal. Chem. **57**, 1884 (1985)
4.22 F. Chandezon, B. Huber, C. Ristori: Rev. Sci. Instrum. **65**, 3344 (1994)
4.23 R. Weinkauf, K. Walter, C. Weickhardt, U. Boesl, E.W. Schlag: Z. Naturforsch. A **44**, 1219 (1989)
4.24 P. Piseri, S. Iannotta, P. Milani: Int. J. Mass Spectrom. Ion Process. **153**, 23 (1996)
4.25 M. Meron: Nucl. Instrum. Methods A **330**, 259 (1993)
4.26 B.A. Mamyrin, V.I. Karataev, D.V. Schmikk, V.A. Zagulin: Sov. Phys.-JETP **37**, 45 (1973)
4.27 T.I. Wang, C.W. Chu, H.M. Hung, G.S. Kuo, C.C. Han: Rev. Sci. Instrum. **65**, 1585 (1994)
4.28 T. Bergmann, T.P. Martin, H. Schaber: Rev. Sci. Instrum. **60**, 792 (1989)
4.29 H. Falter, O.F. Hagena, W. Henkes, H. von Wedel: Int. J. Mass Spectrom. Ion Phys. **4**, 145 (1970)
4.30 O. Hagena: in *Molecular Beams and Low Density Gasdynamics,* P.P. Wegener (ed.), (Dekker, New York, 1974)
4.31 D. Turner, H. Shanks: J. Appl. Phys. **70**, 5385 (1991)

4.32 T.D. Märk: in *Nuclear Physics Concepts in the Study of Atomic Cluster Physics*, R. Schmidt, H.O. Lutz, R. Dreizler (eds.), Lecture Notes in Physics 404, (Springer-Verlag, Berlin, 1992)
4.33 W.A. de Heer: Rev. Mod. Phys. **65**, 611 (1993)
4.34 P. Milani, W.A. de Heer, A Chatelain: Z. Phys. D **19**, 133 (1991)
4.35 W. Henkes, V. Hoffman, F. Mikosch: Rev. Sci. Instrum. **48**, 675 (1977)
4.36 J.L. Wiza: Nucl. Instrum. Methods **162**, 587 (1979)
4.37 R.A. Baragiola: Nucl. Instrum Methods Phys. Res. B **78**, 223 (1993)
4.38 K. Töglhofer, F. Aumayr, H. Kurz, HP. Winter, P. Scheier, T.D. Märk: J. Chem. Phys. **99**, 8254 (1993)
4.39 M. Fallavier: Nucl. Instrum. Methods Phys. Res. B **112**, 72 (1996)
4.40 R.A. Baragiola: Nucl. Instrum. Methods B **88**, 35 (1994)
4.41 U. Even, P.J. de Lange, H.T. Jonkman, and J. Kommandeur: Phys. Rev. Lett. **56**, 965 (1986)
4.42 P.U. Andersson, and J.B.C. Pettersson: Z. Phys. D **41**, 57 (1997)
4.43 U. Zimmermann, N. Malinowski, U. Näher, S. Frank, T.P. Martin: Z. Phys. D **31**, 85 (1994)
4.44 N.R. Daly: Rev. Sci. Instrum. **31**, 264 (1960)
4.45 B.W. Ridley: Nucl. Instrum. Methods **14**, 231 (1961)
4.46 L.A. Dietz, J.C. Sheffield: Rev. Sci. Instrum. **44**, 183 (1973)
4.47 W.R. Gentry: in *Atomic and Molecular Beam Methods*, G. Scoles (ed.), Vol. I, Chap. 3., (Oxford University Press, Oxford, 1988)
4.48 D.J. Auerbach: in *Atomic and Molecular Beam Methods*, G. Scoles (ed.), Vol. I, Chap. 14., (Oxford University Press, Oxford, 1988)
4.49 J.B. Anderson, J.B. Fenn: Phys. Fluids, **8** 780 (1965).
4.50 P. Piseri, A. Li Bassi, P. Milani: Rev. Sci. Instrum. **69**, 1647 (1998)
4.51 G.F. Knoll: *Radiation Detection and Measurement*, (Wiley, New York, 1979).
4.52 M. Ehbrecht, H. Ferkel, F. Huisken: Z. Phys. D **40**, 88 (1997)
4.53 M. Ehbrecht, B. Kohn, F. Huisken, M.A. Laguna, V. Paillard: Phys. Rev. B **56**, 6958 (1997)
4.54 K.H. Homann, J. Traube: Ber. Bunsenges. Phys. Chem. **91**, 833 (1987)
4.55 Bu. Wrenger, K.H. Meiwes-Broer, O. Speer, M.E. Garcia: Phys. Rev. Lett. **79**, 2562 (1997)
4.56 Bu. Wrenger, K.H. Meiwes-Broer: Rev. Sci. Instrum. **68**, 2027 (1997)
4.57 P. Fayet, F. Patthey, H.-V. Roy, Th. Detzel, W.-D. Schneider: Surf. Sci. **269/270**, 1101 (1992)
4.58 P. Fayet, L. Wöste: Surf. Sci. **156**, 134 (1984)
4.59 P. Fayet: Dissertation, EPFL, Lausanne (1987), unpublished
4.60 U. Heiz, F. Vanolli, L. Trento, W.-D. Schneider: Rev. Sci. Instrum. **68**, 1986 (1997)
4.61 Y. Kuk, M.F. Jarrold, P.J. Silverman, J.E. Bower, W.L. Brown: Phys. Rev. B **39**, 11168 (1989)
4.62 K. Broman, C. Felix, H. Brune, W. Harbich, R. Monot, J. Buttet, K. Kern: Science **274**, 956 (1996)
4.63 W. Harbich, F. Meyer, D.M. Lindsay, J. Lignieres, J.C. Rivoal, D. Kreisle: J. Chem. Phys. **93**, 8535 (1990)
4.64 E.W. Becker, K. Bier, W. Bier: Z. Naturforsch. A **16**, 12 (1961)
4.65 V.H. Reis, J.B. Fenn: J. Chem. Phys. **39**, 3240 (1963)
4.66 J. Fernandez De La Mora, B.L. Halpern, J.A. Wilson: J. Fluid Mech. **149**, 217 (1984)
4.67 G.W. Israel, S.K. Friedlander: J. Colloid. Interface Sci. **24**, 330 (1967)
4.68 E. Barborini et al.: Chem. Phys. Lett. **300**, 633 (1999)
4.69 E. Barborini, P. Piseri, S. Mutti, P. Milani, F. Biasioli, S. Iannotta, S. Gialanella: Nanostruct. Mater. **10**, 1023 (1998)

Chapter 5

5.1 J.E. Sundgren: in *Diamond and Diamond-like Films and Coatings*, R.E. Clausing, L.L. Horton, J.C. Angus, P. Koidl (eds.), NATO ASI Series B 266, (Plenum Press, New York, 1991)

5.2 I. Yamada: in *Application of Accelerators in Research and Industry*, J.L. Duggan, I.L. Morgan (eds.), AIP Conf. Proc. **392**, 479 (1997)

5.3 J. Matsuo, N. Toyoda, I. Yamada: J. Vac. Sci. Technol. B **14**, 3951 (1996)

5.4 R.A. Gottscho, M.E. Barone, J.M. Cook: MRS Bull. **21**, 38 (1996), and references therein.

5.5 S. Kashu, E. Fuchita, T. Manabe, C. Hayashi: Jpn. J. Appl. Phys. **23**, L910 (1984)

5.6 T. Takagi, I. Yamada, A. Sasaki: Thin Solid Films **39**, 207 (1976)

5.7 J.Y. Tsao, E. Chason, K.M. Horn, D.K. Brice, S.T. Picraux: Nucl. Instrum. Methods Phys. Res. B **39**, 72 (1989)

5.8 K.-H. Müller: J. Appl. Phys. **61**, 2516 (1987)

5.9 C.L. Cleveland, U. Landman: Science **257**, 355 (1992)

5.10 R. Biswas, G.S. Grest, C.M. Soukoulis: Phys. Rev. B **38**, 8154 (1988)

5.11 I. Kwon, R. Biswas, G.S. Grest, C.M. Soukoulis: Phys. Rev. B **41**, 3678 (1990)

5.12 H. Hsie, R.S. Averback, H. Sellers, C.P. Flynn: Phys. Rev. B **45**, 4417 (1992)

5.13 H. Haberland, Z. Insepov, M. Moseler: Phys. Rev. B **51**, 11061 (1995)

5.14 P. Melinon, G. Fuchs, B. Cabaud, A. Hoareau, P. Jensen, V. Paillard, M. Treilleux: J. Phys. I **3**, 1585 (1993)

5.15 G.L. Kellog: Surf. Sci. Rep. **21**, (1994)

5.16 S.J. Carrol, P. Weibel, B. von Issendorf, L. Kuipers, R.E. Palmer: J. Phys.: Condens. Matter **8**, L617 (1996)

5.17 Q. Zhong, D. Innis: Surf. Sci. **290**, L688 (1993)

5.18 L. Bardotti, P. Jensen, A. Hoareau, M. Treilleux, B. Cabaud: Phys. Rev. Lett. **74**, 23 (1995)

5.19 L. Bardotti, P. Jensen, A. Hoareau, M. Treilleux, B. Chabaud, A. Perez, F. Cadete Santos Aires: Surf. Sci. **367**, 276 (1996)

5.20 A. Perez, P. Melinon, V. Dupuis, P. Jensen, B. Prevel, J. Tuallion, L. Bardotti, C. Martet, M. Treuillex, M. Broyer, M. Pellarin, J.L. Vaille, B. Palpant and J. Lerme: J. Phys. D: Appl. Phys. **30**, 709 (1997).

5.21 J.F. Roux, B. Cabaud, G. Fuchs, D. Guillot, A. Hoareau, P. Melinon: Appl. Phys. Lett., **64**, 10 (1994).

5.22 I. Moullet: Surf. Sci. **331**, 697 (1995)

5.23 S.J. Carroll, K. Seeger, R.E. Palmer: Appl. Phys. Lett., **72**, 305 (1998)

5.24 P. Melinon, V. Paillard, V. Dupuis, A. Perez, P. Jensen, A. Hoareau, M. Broyer, J.L. Vialle, M. Pellarin, B. Baguenard, J. Lerme: Int. J. Mod. Phys. B**9**, 339 (1995)

5.25 P. Jensen, A.L. Barabasi, H. Larralde, S. Havlin, H.E. Stanley: Nature **22**, 368 (1994)

5.26 P. Jensen, A.L. Barabasi, H. Larralde, S. Halvin, H.E. Stanley: Phys Rev B**50**, 15316 (1994)

5.27 G.S. Bales, D.C. Chrzan: Phys. Rev. B**50**, 6057 (1994)

5.28 C. Ratsch, A. Zewail, P. Similauer, D.D. Vvdensky: Phys. Rev. Lett. **72**, 3194 (1994)

5.29 M.C. Bartelt, J.W. Evans: Surf. Sci. **298**, 421 (1993)

5.30 J.M. When, S.L. Chang, J.W. Burnett, J.W. Evansand, A. P. Thiel: Phys. Rev. Lett. **73**, 2591 (1994)

5.31 A. Masson, J.J. Metois, R. Kern: Surf. Sci. **27**, 463 (1971)

5.32 I.M. Goldby, L. Kuipers, B. von Issendorff, R.E. Palmer: Appl. Phys. Lett. **69**, 2819 (1996)
5.33 P. Blandin, C. Massobrio, P. Ballone: Phys. Rev. Lett. **72**, 3072 (1994)
5.34 G.M. Francis, L. Kuipers, J.R.A. Cleaver, R.E. Palmer: J. Appl. Phys. **79**, 2942 (1996)
5.35 K. Bromann, H. Brune, C. Felix, W. Harbich, R. Monot, J. Buttet, K. Kern: Surf. Sci. **377**, 1051 (1997)
5.36 G. Vandoni, C. Felix, R. Monot, J. Buttet, W. Harbich: Chem. Phys. Lett. **229**, 51 (1994)
5.37 H. Röder, H. Brune, K. Kern: Phys. Rev. Lett. **73**, 2143 (1994)
5.38 H. Gleiter: Prog. Mater. Sci. **33**, 223 (1989)
5.39 C.L. Chien: in *Science and Technology of Nanostructured Magnetic Materials*, G.C. Hadjipanayis, G.A. Prinz (eds.), (Plenum Press, New York, 1991)
5.40 G.S. Bales, R. Bruinsma, E.A. Eklund, R.P.U. Karunasiri, J. Rudnick, A. Zangwill: Science **249**, 264 (1990)
5.41 B. Fultz, C.C. Ahn, E.E. Alp, W. Sturhahn, T.S. Toellner: Phys. Rev. Lett. **79**, 937 (1997)
5.42 T. Junno, S. Anand, K. Deppert, L. Montelius, L. Samuelson: Appl. Phys. Lett. **66**, 3295 (1995)
5.43 S.M. Prokes: in *Nanomaterials: Synthesis, Properties and Applications*, A.S. Edelstein, R.C. Cammarata (eds.), (IOP Publishing, Bristol, 1996)
5.44 Y. Maeda: Phys. Rev. B **51**, 1658 (1995)
5.45 L.Y. Canham: Appl. Phys. Lett. **57**, 1046 (1990)
5.46 R. Tsu, H. Shen, M. Dutta: Appl. Phys. Lett. **60**, 112 (1992)
5.47 S. Hayashi, K. Yamamoto: J. Lumin. **70**, 352 (1996)
5.48 M.C. Klein, F. Hache, D. Ricard, C. Flytzanis: Phys. Rev. B **42**, 11123 (1990)
5.49 S. Hayashi, M. Ito, H. Kanamori: Solid State Commun. **44**, 75 (1982)
5.50 C.E. Bottani, C. Mantini, P. Milani, M. Manfredini, A. Stella, P. Tognini, P. Cheyssac, R. Kofman: Appl. Phys. Lett. **69**, 2409 (1996)
5.51 M. Zacharias, J. Blaesing, J. Christen, P. Veit, B. Dietrich, D. Bimberg: Superlattices Microstruct. **18**, 139 (1995)
5.52 M. Fujii, S. Hayashi, K. Yamamoto: Jpn. J. Appl. Phys. **44**, 687 (1991)
5.53 J. Fortner, R.Q. Yu, J.S. Lannin: Phys. Rev. B **42**, 7610 (1990)
5.54 Y. Kanemitsu, H. Uto, Y. Masumoto, Y. Maeda: Appl. Phys. Lett. **61**, 2187 (1992)
5.55 A.K. Dutta: Appl. Phys. Lett. **68**, 1189 (1996)
5.56 D.C. Paine, C. Caragianis, T.Y. Kim, Y. Shigesato, T. Ishahara: Appl. Phys. Lett. **62**, 2842 (1993)
5.57 M. Zacharias, R. Weigand, B. Dietrich, F. Stolze, J. Blaesing, P. Veit, T. Druesedau, J. Christen: J. Appl. Phys. **81**, 2384 (1997)
5.58 C. Caragianis-Broadbridge, J.M. Blaser, D.C. Paine: J. Appl. Phys. **82**, 1626 (1997)
5.59 T. Kobayashi, T. Endoh, H. Fukuda, S. Nomura, A. Sakai, Y. Ueda: Appl. Phys. Lett. **71**, 1195 (1997)
5.60 H. Miguez, V. Fornes, F. Meseguer, F. Marquez, C. Lopez: Appl. Phys. Lett. **69**, 2347 (1996)
5.61 J.R. Heath, J.J. Shiang, A.P. Alivisatos: J. Chem. Phys. **101**, 1607 (1994)
5.62 J.G. Zhu, C.W. White, J.D. Budai, S.P. Withrow, Y. Chen: J. Appl. Phys. **78**, 4386 (1995)
5.63 S. Sato, S. Nozaki, H. Morisaki, M. Iwase: Appl. Phys. Lett. **66**, 3176 (1995)
5.64 A. Gonzalez-Hernandez, G.H. Azarbayejani, R. Tsu, F.H. Pollak: Appl. Phys. Lett. **47**, 1350 (1985)
5.65 M. Wakaki, M. Iwase, Y. Show, K. Koyama, S. Sato, S. Nozaki, H. Morisaki: Physica B **219&220**, 535 (1996)

5.66 M. Ehbrecht, H. Ferkel, V.V. Smirnov, O.M. Stelmakh, W. Zhang, F. Huisken: Rev. Sci. Instrum. **66**, 3833 (1995)
5.67 M. Ehbrecht, H. Ferkel, F. Huisken: Z. Phys. D **40**, 88 (1997)
5.68 M. Ehbrecht, B. Kohn, F. Huisken, M.A. Laguna, V. Paillard: Phys. Rev. B **56**, 6958 (1997)
5.69 P.M. Fauchet, J.H. Campbell: Crit. Rev. Solid State Mater. Sci. **14**, S79 (1988)
5.70 H. Richter, Z.P. Wang, L. Ley: Solid State Commun. **39**, 625 (1981)
5.71 Y. Guyot, B. Champagnon, M. Boudeulle, P. Melinon, B. Prevel, V. Dupuis, A. Perez: Thin Solid Films **297**, 188 (1997)
5.72 P. Melinon, P. Keghelian, B. Prevel, A. Perez, G. Guiraud, J. LeBrusq, J. Lerme, M. Pellarin, M. Broyer: J. Chem. Phys. **107**, 10278 (1997)
5.73 P. Milani, M. Ferretti, P. Piseri, C. Bottani, A. Ferrari, A. Li Bassi, G. Guizzetti, M. Patrini: J. Appl. Phys. **82**, 5793 (1997)
5.74 P. Milani, P. Piseri: *Application of Accelerators in Research and Industry*, J.L. Duggan and I.L. Morgan (eds.), AIP Conf. Proc. **392**, 495 (1997)
5.75 A. Canning, G. Galli, J. Kim: Phys. Rev. Lett. **78**, 4442 (1997)
5.76 H. Sjostrom, S. Stafstrom, M. Roman, J.-F. Sundgren: Phys. Rev. Lett. **75**, 1336 (1995)
5.77 G.A.J. Amaratunga, M. Chowalla, C.J. Kiely, I. Alexandrou, R. Aharonov, R.M. Devenish: Nature, **383**, 321 (1996)
5.78 K.W.R. Gilkes, H.S. Sands, D.N. Batchelder, J. Robertson, W.I. Milne: Appl. Phys. Lett. **70**, 1980 (1997)
5.79 G. Benedek, L. Colombo: Mater. Sci. Forum, **232**, 247 (1996)
5.80 P. Milani, C.E. Bottani: *Vibrational Properties of Mesoscopic Systems*, in Handbook of Nanostructured Materials and Nanotechnology, H. Nalwa (ed.), (Academic), in press. (1999)
5.81 M. Grimsditch: J. Phys. C **16**, L143 (1983)
5.82 X. Jiang, J. V. Harzer, B. Hillebrands, Ch. Wild and P. Koidl: Appl. Phys. Lett. **59**, 1055 (1991)
5.83 X. Jiang, J. W. Zou, K. Reichelt and P. Grünberg: J. Appl. Phys. **66**, 4729 (1989)
5.84 D. Fioretto, G. Carlotti, G. Socino, S. Modesti, C. Cepek, L. Giovannini, O. Donzelli and F. Nizzoli: Phys. Rev. B **52**, R8707 (1995)
5.85 M. Manfredini, C. E. Bottani and P. Milani: Chem. Phys. Lett. **226**, 600 (1994)
5.86 T.A. Friedmann, J.P. Sullivan, J.A. Knapp, D.R. Tallant, D.M. Follstaedt, D.L. Medlin, P.B. Mirkarimi: Appl. Phys. Lett. **71**, 3820 (1997)
5.87 C.E. Bottani, A.C. Ferrari, A. Li Bassi, P. Milani, P. Piseri: Europhys. Lett. **42**, 431 (1998)
5.88 A. Griffin: Rev. Mod. Phys. **40**, 167 (1968)
5.89 O.L. Blakslee, D.G. Proctor, E.J. Seldin, G.B. Spence, T. Weng: J. Appl. Phys. **41**, 3373 (1970)
5.90 E.J. Seldin, C.W. Nezbeda: J. Appl. Phys. **41**, 3389 (1970)
5.91 G. Fuchs, C. Montadon, M. Treilleux, J. Dumas, B. Cabaud, P. Melinon, A. Hoareau: J. Phys. D; Appl. Phys. **26**, 1114 (1993)
5.92 G. Fuchs, P. Melinon, F. Santos Aires, M. Treilleux, B. Cabaud, A. Hoareau: Phys. Rev. B **44**, 3926 (1991)
5.93 J. Tuaillon, V. Dupuis, P. Melinon, B. Prevel, M. Treilleux, A. Perez, M. Pellarin, J.L. Vialle, M. Broyer: Philos. Mag. A **76**, 493 (1997)
5.94 V. Dupuis, J.P. Perez, J. Tuaillon, V. Paillard, P. Melinon, A. Perez, B. Barbara, L. Thomas, S. Fayeulle, J.M. Gray: J. Appl. Phys. **76**, 6676 (1994)

5.95 D.A. Easthman, Y. Qiang, T.H. Maddock, J. Kraft, J.P. Schille, G.S. Thompson, H. Haberland: J. Phys., Condens. Matter **9**, L497 (1997)
5.96 G.F. Hohl, T. Hihara, M. Sakurai, T.J. Konno, K. Sumiyama, F. Hensel, K. Suzuki: Appl. Phys. Lett. **66**, 385 (1995)
5.97 G.F. Hohl, T. Hihara, M. Sakurai, K. Sumiyama, F. Hensel, K. Suzuki, J.A. Becker: Mater. Sci. Eng. A **217/218**, 291 (1996)
5.98 S. Yamamuro, M. Sakurai, T.J. Konno, K. Sumiyama, K. Suzuki: in *Similarities and Differences between Atomic Nuclei and Clusters*, Y. Abe, I. Arai, S.M. Lee, K. Yabana (eds.), AIP Conf. Proc. **416**, 491 (1998)
5.99 I.M. Billas, A. Chatelain, W.A. deHeer: Science **265**, 1682 (1994)
5.100 M.W. Mathew, R.J. Beuhler, M. Ledbetter, L. Friedman: J. Phys. Chem. **90**, 3152 (1986)
5.101 R.J. Beuheler, L. Friedman: J. Phys. **50**, C2-127 (1989)
5.102 I. Yamada, J. Matsuo: Mat. Res. Soc. Symp. Proc. **427**, 265 (1996)
5.103 J. Gspann: Nucl. Instrum. Methods Phys. Res. B **112**, 86 (1996)
5.104 J. Matsuo, N. Toyoda, M. Akizuki, I Yamada: Nucl. Instrum. Methods Phys. Res. B **121**, 459 (1997)
5.105 H. Kitani, N. Toyoda, J. Matsuo, I Yamada: Nucl. Instrum. Methods Phys. Res. B **121**, 489 (1997)
5.106 D. Takeuci, K. Fukushima, J. Matsuo, I Yamada: Nucl. Instrum. Methods Phys. Res. B **121**, 493 (1997)
5.107 T. Seki, T. Kaneko, D. Takeuchi, T. Aoki, J. Matsuo, Z. Insepov, I Yamada: Nucl. Instrum. Methods Phys. Res. B **121**, 498 (1997)
5.108 Z. Insepov, I. Yamada, M. Sosnowski: J. Vac. Sci. Technol. A **15**, 981 (1997)
5.109 C. Ascheron: private communication
5.110 P.R.W. Henkes, B. Krevet: J. Vac. Sci. Technol. A **13**, 2133 (1995)
5.111 M. Tanomura, D. Takeuchi, J. Matsuo, G.H. Takaoka, I Yamada: Nucl. Instrum. Methods Phys. Res. B **121**, 480 (1997)
5.112 M Döbeli, F. Ames, R.M. Ender, M. Suter, H.A. Synal, D. Vetterli: Nucl. Instrum. Methods Phys. Res. B **106**, 43 (1995)
5.113 J.A. Northby, T. Jiang, G.H. Takaoka, I. Yamada, W.L. Brown, M. Sosnowski: Nucl. Instrum. Methods Phys. Res. B **74**, 336 (1993)
5.114 I. Yamada, W.L. Brown, J.A. Northby, M. Sosnowski: Nucl. Instrum. Methods Phys. Res. B **79**, 223 (1993)
5.115 P.R.W. Henkes, R. Klingelhöfer: J. Phys. **50**, C2-127 (1989)
5.116 P.R.W. Henkes: Rev. Sci. Instrum. **61**, 360 (1990)
5.117 J. Gspann: in *Large Clusters of Atoms and Molecules*, T.P. Martin (ed.), Nato ASI Series E 313, (Kluwer, Dordrecht, 1996)
5.118 T. Takagi, I. Yamada, A. Sasaki: J. Vac. Sci. Technol. **12**, 1128 (1975)
5.119 I. Yamada, T. Takagi: Thin Solid Films **80**, 105 (1981)
5.120 J.E. Greene, S.A. Barnett, J.-E. Sundgren, A. Rockett: in *Plasma-Surface Interactions and Processing of Materials*, O. Auciello, A. Gras-Marti, D.L. Flamm (eds.), NATO ASI, Ser. E, 176, (Kluwer Academic, Boston, 1990)
5.121 I. Yamada, H. Takaoka, H. Inokawa, H. Usui, S.C. Cheng, T. Takagi: Thin Solid Films **92**, 137 (1982)
5.122 F.K. Urban III, S.W. Feng, J.J. Nainaparampil: Appl. Phys. Lett. **61**, 180 (1992)
5.123 S. Sato, S. Nozaki, H. Morisaki: J. Appl. Phys. **81**, 1518 (1997)
5.124 I Yamada: Nucl. Instrum. Methods Phys. Res. B **55**, 544 (1991), and references therein.
5.125 H. Ito, Y. Minowa, T. Ina, K. Yamanishi, S. Yasunaga: Rev. Sci. Instrum. **61**, 604 (1990)

5.126 K. Yamanishi, Y. Hashimoto, H. Tsukazaki: Nucl. Instrum. Methods Phys. Res. B **99**, 233 (1995)
5.127 T. Hihara, K. Sumiyama, M. Sakurai, H. Onodera, K. Wakoh, K. Suzuki: J. Phys. Soc. Jpn. **66**, 1450 (1997)
5.128 O.F. Hagena: Rev. Sci. Instrum. **63**, 2374 (1992)
5.129 Y. Franghiadakis, P. Tzanetakis: J. Appl. Phys. **68**, 2433 (1990)
5.130 W.L. Brown, M.F. Jarrold, R.L. McEachern, M. Sosnowski, G. Takaoka, H. Usui, I. Yamada: Nucl. Instrum. Methods Phys. Res. B **59/60**, 182 (1991)
5.131 D. Turner, H. Shanks: J. Appl. Phys. **70**, 5385 (1991)
5.132 O.F. Hagena, G. Knop, R. Fromknecht, G. Linker: J. Vac. Sci. Technol. A **12**, 282 (1994)
5.133 P. Gatz, O.F. Hagena: Appl. Surf. Sci. **91**, 169 (1995)
5.134 P. Gatz, O.F. Hagena: J. Vac. Sci. Technol. A **13**, 2128 (1995)
5.135 J. Gspann: Nucl. Instrum. Methods Phys. Res. B **37/38**, 775 (1989)
5.136 H. Haberland, M. Karrais, M. Mall, Y. Thurner: J. Vac. Sci. Technol. A **10**, 3266 (1992)
5.137 H. Haberland, M. Mall, M. Moseler, Y. Qiang, T. Reiners, Y. Thurner: J. Vac. Sci. Technol. A **12**, 2925 (1994)
5.138 Y. Qiang, Y. Thurner, T. Reiners, O. Rattunde, H. Haberland: E-MRS 97, to be published.

Chapter 6

6.1 A.J. Nozik, O.I. Mićić: MRS Bull. **23**, 24 (1998)
6.2 I. Yamada: J. Matsuo: MRS Proc. **427**, 265 (1996)
6.3 I. Yamada: Nucl. Instrum. Methods Phys. Res. B **99**, 240 (1995)
6.4 P.R.W. Henkes, B. Krevet: J. Vac. Sci. Technol. **13**, 2133 (1995)
6.5 J. Gspann: Nucl. Instrum. Methods Phys. Res. B **112**, 86 (1996)
6.6 I. Yamada: *Advances in Ion and Laser Beam Technology: Achievements of Japanese Government and University Projects*, Symp: Atomistic Mechanisms in Beam Synthesis and Irradiation of Materials, MRS Fall Meeting, Boston (1997)
6.7 W.K. Chu, Y.P. Li, J.R. Liu, J.Z. Wu, S.C. Tidrow, N. Toyoda, J. Matsuo, I. Yamada: Appl. Phys. Lett. **72**, 246 (1998)
6.8 D. Takeuchi, N. Shimada, J. Matsuo, I. Yamada: Nucl. Instrum. Methods Phys. Res. B **121**, 345 (1997)
6.9 J. Matsuo, I. Yamada: Mater. Sci. Semicond. Process. **1**, 27 (1998)
6.10 P. Gatz, O.F. Hagena: Appl. Surf. Sci. **91**, 169 (1995)
6.11 J. Matsuo, H. Abe, G.H. Takaoka, I Yamada: Nucl. Instr. and Meth. Phys. Res. B **99**, 244 (1995)
6.12 D. Eres, D.H. Lowndes, J.Z. Tischler: Appl. Phys. Lett. **55**, 1008 (1989)
6.13 E. Barborini, P. Piseri. S. Mutti, P. Milani, F. Biasioli, S. Iannotta, S. Gialanella: NanoStructur. Mater. **10**, 1023 (1998)
6.14 S. Tiwari, F. Rana, K. Chan, L. Shi, H. Hanafi: Appl. Phys. Lett. **69**, 1232 91996)
6.15 see, for example: *Functionally Gradient Materials*, MRS Bull., XX, No. 1 (1995); P. Milani et al.: Europ. Phys. J. D, in press (1999)
6.16 T. Toccoli et al.: Phil. Mag. B (1999), in press
6.17 G. Ciullo, M. Moratti, T. Toccoli, S. Iannotta: Phil. Mag. B (1999), in press

Printing: Saladruck, Berlin
Binding: Buchbinderei Lüderitz & Bauer, Berlin